HELLCATS

"Only a little way into *Hellcats*, the reader may well find himself wondering when this story will make it to the big screen. Fast-paced yet attentive to d~~etails such as the baffling~~ ~~~~ ~~~~ ~~~~ eaucracy during World ~~~~ ~~~~ ~~~~ account and a real page ~~~~ ~~~~ ~~~~ accounts of undersea c~~~~ ~~~~ ~~~~ combat reports. . . . [S ~~~~ ~~~~ ~~~~ l fire his adrenaline an~~~~ ~~~~ ~~~~ *n WWII*

"A worthwhile ~~~~ ~~~~ ~~~~ accounts for the genera~~~~ ~~~~ ~~~~ picking of the subma~~~~ ~~~~ ~~~~ ble and moving . . . de~~~~ ~~~~ ~~~~ *Booklist*

"Sasgen vividl~~~~ ~~~~ ~~~~ . . . well-written, engag~~~~ ~~~~ ~~~~ bmarine war against Ja~~~~ ~~~~ ~~~~ *History*

Other Books by Peter Sasgen

Stalking the Red Bear: The True Story of a U.S. Cold War Submarine's Covert Operations Against the Soviet Union

Red Scorpion: The War Patrols of the USS Rasher

War Plan Red (novel)

Red Shark (novel)

HELLCATS

The Epic Story
of World War II's
Most Daring
Submarine Raid

PETER SASGEN

NAL
CALIBER

New American Library
Published by New American Library, a division of
Penguin Group (USA) Inc., 375 Hudson Street,
New York, New York 10014, USA
Penguin Group (Canada), 90 Eglinton Avenue East, Suite 700, Toronto,
Ontario M4P 2Y3, Canada (a division of Pearson Penguin Canada Inc.)
Penguin Books Ltd., 80 Strand, London WC2R 0RL, England
Penguin Ireland, 25 St. Stephen's Green, Dublin 2,
Ireland (a division of Penguin Books Ltd.)
Penguin Group (Australia), 250 Camberwell Road, Camberwell, Victoria 3124,
Australia (a division of Pearson Australia Group Pty. Ltd.)
Penguin Books India Pvt. Ltd., 11 Community Centre, Panchsheel Park,
New Delhi - 110 017, India
Penguin Group (NZ), 67 Apollo Drive, Rosedale, Auckland 0632,
New Zealand (a division of Pearson New Zealand Ltd.)
Penguin Books (South Africa) (Pty.) Ltd., 24 Sturdee Avenue,
Rosebank, Johannesburg 2196, South Africa

Penguin Books Ltd., Registered Offices:
80 Strand, London WC2R 0RL, England

Published by NAL Caliber, a division of Penguin Group (USA) Inc. Previously published in an NAL
Caliber hardcover edition.

First NAL Caliber Trade Paperback Printing, November 2011
10 9 8 7 6 5 4 3 2 1

Copyright © Peter Sasgen, 2010
Maps by Karen Sasgen
All rights reserved

NAL CALIBER and the "C" logo are trademarks of Penguin Group (USA) Inc.

NAL Caliber Trade Paperback ISBN: 978-0-451-23485-8

THE LIBRARY OF CONGRESS HAS CATALOGUED THE HARDCOVER EDITION AS FOLLOWS:
Sasgen, Peter T., 1941–
    Hellcats: the epic story of World War II's most daring submarine raid/Peter Sasgen.
      p. cm.
    Includes bibliographical references and index.
    ISBN 978-0-451-23136-9
    1. World War, 1939–1945—Naval operations—Submarine. 2. World War, 1939–1945—Naval
operations, American. 3. World War, 1939–1945—Campaigns—Japan, Sea of. 4. Operation
Barney, 1945. 5. Lockwood, Charles A., 1890–1967. 6. Sonar—History—20th century. I. Title.
    D783.S37 2010
    940.54'25—dc22   2010028766

Set in Minion
Designed by Ginger Legato

Printed in the United States of America

To the men and the families

of the USS *Bonefish* (SS-223)

CONTENTS

Part Three: Operation Barney

The Americans had not . . . made any great sacrifices of blood. They would certainly not withstand a great trial by fire, for their fighting qualities were low. In general, no such thing as an American people existed as a unit; they were nothing but a mass of immigrants from many nations and races.

—Adolf Hitler, from *Inside the Third Reich* by Albert Speer

To our good and loyal subjects . . . We declared war on America and Britain out of our sincere desire to ensure Japan's self-preservation and the stabilization of East Asia, it being far from our thought either to infringe upon the sovereignty of other nations or to embark upon territorial aggrandizement. Despite [this] the war situation has developed not necessarily to Japan's advantage.

—Excerpt from Emperor Hirohito's surrender address to the Japanese people, August 15, 1945

It is to the everlasting honor and glory of our submarine personnel that they never failed us in our days of great peril.

—Chester W. Nimitz, Fleet Admiral, United States Navy

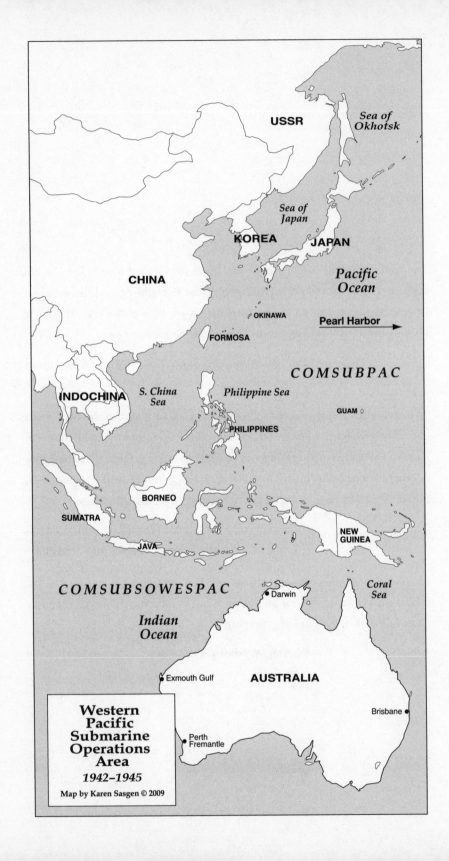

Western
Pacific
Submarine
Operations
Area
1942–1945

Map by Karen Sasgen © 2009

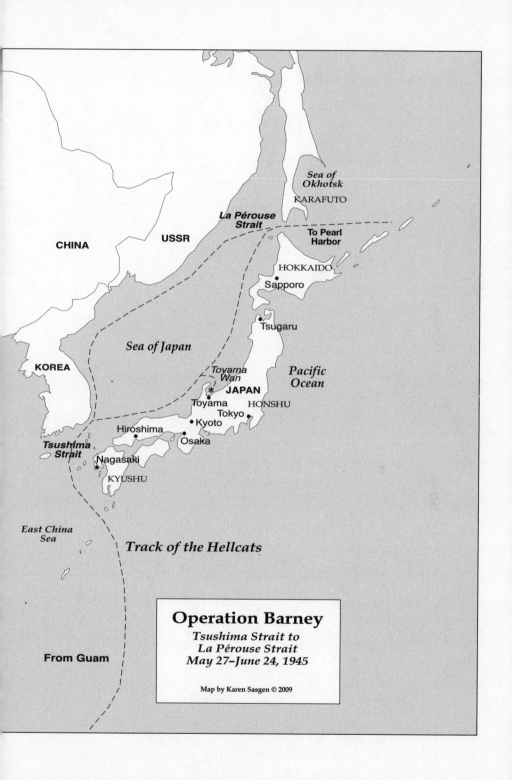

Sea of
Okhotsk

KARAFUTO

CHINA   USSR

La Pérouse
Strait

To Pearl
Harbor

HOKKAIDO
Sapporo

Tsugaru

Sea of Japan   Pacific
Ocean

KOREA

Toyama
Wan   *JAPAN
Toyama
Tokyo
Kyoto   HONSHU
Hiroshima
Osaka

Tsushima
Strait

Nagasaki

KYUSHU

East China
Sea   Track of the Hellcats

Operation Barney
*Tsushima Strait to
La Pérouse Strait
May 27–June 24, 1945*

From Guam

Map by Karen Sasgen © 2009

May 27, 1945. Diesel engines rumbling impatiently, a task force of submarines lay moored in Apra Harbor, Guam. Fueled, provisioned, torpedoes loaded, the subs and their crews were ready to sail. As the late-afternoon hour for their departure approached, submarine force commander Vice Admiral Charles A. Lockwood and his staff gathered to see the subs off on their mission.

With great anticipation and excitement, they watched the subs, their diesels rolling to a deep thrumming pitch, cast off lines and clear their moorings. As the task force formed up and moved slowly toward the open sea, Lockwood returned departing salutes and waved good-bye.

Earlier, Lockwood had thought to say something beyond the rote custom of "good luck" and "good hunting," something that the submariners could draw strength from in difficult moments. But he had seen the steady burn of self-confidence in the faces of the departing skippers and knew that the time for speeches was over. The mission was in their hands now. And in their hands it would either succeed or fail; there would be no middle ground.

Lockwood felt an all too familiar pang of apprehension. There were so many unknowns. So many factors beyond the submariners' control. Anything could go wrong. For one, the enemy was unpredictable. For another, the secret sonar device aboard those subs, which had made the mission possible and on which so much was riding for its success, was neither perfect nor foolproof.

Lockwood had never felt more alone than he did at that moment. Yet, at the same time he felt more connected than ever to the almost eight hundred men departing on what might prove to be a doomed mission. Two years had passed since the loss of the storied submarine *Wahoo* and her gallant crew, which had inspired the mission. During that time more submarines had been lost and Lockwood feared that the mission on which he was sending the task force would result in further losses. A picture formed in his mind, one that had haunted him day and night since the war began: a submarine, destroyed by depth charges, plunging into a black void.

Not everyone in the submarine force shared Lockwood's ironbound confidence in the secret electronic weapon aboard those subs, nor his unshakable belief that the mission, if it succeeded, would help end the war. Officers whose judgment he respected had told him that it was a suicide mission. But there were others, especially the scientists who had worked like demons to perfect the secret weapon, who believed it would work and that the mission would succeed. It was, Lockwood reflected, too late to alter plans or have second thoughts.

As the submarines slowly threaded their way out of the harbor to disappear from view into the shimmering Pacific Ocean, Lockwood could only wait for the first reports from the sea to the north where those subs and the ones to follow would fight one of the most daring—and dangerous—submarine battles of World War II.

During the closing days of World War II in the Pacific, nine United States Navy submarines penetrated a curtain of minefields guarding the Sea of Japan to launch a surprise attack on the remnants of the empire's merchant marine lifeline. Known collectively as the Hellcats, the nine submarines were on a mission to destroy that lifeline and hasten Japan's collapse and surrender. The Hellcats' torpedoes sank more than a score of ships and, with them, thousands of tons of the imported food and matériel Japan needed to continue fighting. The Hellcats' mission, code-named Operation Barney, was the most daring submarine raid of all time. *Hellcats* tells the story of how they did it, what they accomplished, and the price they paid for their success.

Operation Barney was fraught with danger. No one, least of all the submariners themselves, knew how it would turn out, given that its success—or failure—would depend in large measure on an unproven secret weapon designed to locate submerged antisubmarine mines. Skeptics thought that Barney, launched just a few weeks before Japan surrendered on August 15, 1945, was nothing more than a technical exercise, if not a stunt to grab headlines for a sub force that, despite its astounding combat record, had operated mostly in the shadows. Among the submariners tapped for the mission were those who believed that their chances of surviving an underwater encounter with a minefield were razor thin. All they knew for certain was that regardless of the outcome it would take extraordinary courage and skill to execute Operation Barney. Indeed, the risks involved *were* great, so great, in fact, that one of the nine raiding submarines did not return from the mission. Nevertheless, just as the war was ending, a second wave of seven submarines followed the pioneering Hellcats into the Sea of Japan (not to be confused with the Inland Sea) to finish the job the Hellcats had started. The submariners who participated in these missions agreed on one thing: If they could deliver a knockout punch to the Japanese it might bring a quick end to the war without the need for a costly invasion of mainland Japan, thus saving countless American lives, if not their own.

No one doubted that Operation Barney was a bold and daring enterprise. Brave and dedicated sailors aboard the Hellcat submarines had risked their lives to carry it out. Scores of scientists and naval officers had worked at a feverish pace to develop and perfect the secret mine-detecting sonar equipment that allowed the raiding subs to break into the Sea of Japan. Mission accomplished, Operation Barney was hailed as a great tactical success that had exceeded expectations. Even so, the loss of a Hellcat sub, the USS *Bonefish* (SS-223) and her crew of eighty-five men was a blow to the submarine force, and, for the families of the men who perished, a crushing tragedy.

The eight surviving Hellcats returned from the Sea of Japan to the heroes' welcome they deserved. After all the speeches and awarding of medals for valor, the war in the Pacific ended with the dropping of atomic bombs and the Soviet Union's declaration of war on Japan. In the excitement and celebration of victory, information about the fate of the *Bone-*

fish and her crew was virtually nonexistent. Rumors circulating among the families of the missing men said that part of the crew had survived her loss. If that was true, they asked, how many had survived and where were they? For months the families clung to the hope that somehow their loved ones might be found alive in liberated POW camps. While the families waited to learn the fate of the missing *Bonefish* crew, reports that men held prisoner for years had been freed kept hopes alive among the families that their men would be found too. Sadly, those hopes died when the Navy announced in early 1946 that all of the American prisoners held by the Japanese had been accounted for and that there were no survivors from the *Bonefish* among them.

The loss of the *Bonefish*, coming as it did so late in the war, later fueled the controversies that arose over the execution and timing of Operation Barney itself. Why, some family members asked, did the Navy undertake such a dangerous mission only ten weeks before the war ended? In light of the atomic bomb and its effect on Japan's surrender, it appeared to some people that Operation Barney hardly seemed worth the risk, for it had had no measurable effect on an already defeated enemy nor influence on Japan's decision to surrender. Others questioned the use of an unproven sonar system to locate antisubmarine mines, which, given the danger such mines posed, some family members believed had sunk the *Bonefish*. Still others questioned the true purpose of Barney, saying it was a costly makework operation designed to keep the sub force occupied.

Aside from the preceding issues raised by the *Bonefish* families and by critics within the Navy, there is at the heart of Operation Barney a point of debate that overrides all the other controversies and that poses an important question that needs an answer: Was the need to avenge the loss of a storied U.S. submarine in the Sea of Japan early in the war, a submarine that was skippered by a man considered the greatest submariner of his generation, the driving force behind Operation Barney? *Hellcats* will attempt to answer that question.

Even though Operation Barney unfolded more than sixty-five years ago and is now largely forgotten, a careful perusal of declassified documents and correspondence provides a compelling account of how the mission was planned and executed. A close reading of the patrol reports of the Hellcat submarines themselves tells a thrilling story of high adventure. *Hellcats*, then, is the story of how men facing long odds against their survival overcame fear, gloried in triumph over the enemy, and kept alive their hopes and dreams for the day when the war would end.

The personal letters of Commander Lawrence Lott Edge, who perished aboard the *Bonefish* during Operation Barney, are replete with passages that touch on these matters. In particular, his letters to his wife, Sarah, tell a heart-wrenching story of love and loss. They contain revealing insights into his state of mind, personal feelings about the war, command at sea, devotion to duty, and, most poignantly, his yearning to survive the war to return home to Sarah and their daughter. These insights are of a kind not often associated with submariners who enjoy a reputation as iron-willed, remarkably self-contained, enduringly fearless individuals. Add to this the correspondence between the Navy and family members desperately searching for information that would help explain the heartbreaking loss they'd suffered, and a deeply human aspect of wartime submarine service emerges. All of this material, in addition to the writings, papers, biography, and memoirs of the Pacific Submarine Force commander, Vice Admiral Charles A. Lockwood, proved a rich trove from which to assemble the *Hellcats* narrative.

I'm especially grateful for the help and encouragement I received from individuals with a personal interest in Operation Barney, in particular those whose fathers either commanded or served in the Hellcat submarines or perished on the mission. These individuals generously provided exceptional material from family archives, large amounts of their time, and, above all, friendship. My own father served in submarines in World War II and I discovered long ago that like our submariner dads, the children of submariners share a special bond. I hasten to add that they are not responsible for any mistakes or errors that appear in *Hellcats*; they are all my doing.

Like Nazi Germany, Japan's world-conquering ambitions ended when that country collapsed. Whether or not Operation Barney contributed to that collapse may never be known for certain. Regardless, nothing can diminish what the Hellcats accomplished during one of the most challenging and dangerous operations of World War II.

A note on geographic place-names. Where possible I've used the names that were commonly in use during World War II and appear in official Navy correspondence and reports, and on maps and nautical charts.

INTRODUCTION

Mention World War II and submarines in the same breath and most people think German U-boats and the Battle of the Atlantic. Beginning in late 1939, until their defeat in the spring of 1943, Hitler's U-boats sank more than 3,500 Allied merchant ships loaded with food, weapons, and raw materials destined for Great Britain and the Soviet Union from ports in North America. This U-boat onslaught came perilously close to defeating Great Britain, already reeling under air attack from Nazi Germany.

Unknown to most people, even to those with more than a passing interest in World War II, is that, like the Battle of the Atlantic, the war waged by American submarines in the Pacific theater against the merchant marine of Axis Japan played a major factor in that country's defeat. It was, as submarine historian Clay Blair remarked, a war within a war. It was so successful that some have argued it was the liquidation of Japan's merchant marine and the blockade of the home islands by U.S. submarines, and not the atomic bomb, that ultimately defeated Japan.

True or not, the facts are impressive.

By early 1945 Japan's ability to import raw materials and food had about reached its end. Imports had been strangled by the U.S. submarine blockade of the home islands. According to the postwar Joint Army Navy Assessment Committee (JANAC), American submarines sank 1,314 Japanese merchant ships totaling 5.3 million tons, not including cargoes, while Japan's merchant marine complement of 122,000 men suffered 116,000 casualties. This was accomplished by a submarine force of roughly 280 submarines, and 50,000 officers and men including staff and support personnel. It was a costly victory: The U.S. lost 52 subs, 41 of them to direct enemy action. Of the approximately 15,000 men who made war patrols, casualties totaled about 3,500.[1] This war of all-out attrition, that is, unrestricted submarine warfare, ranged over eight million square miles of Pacific Ocean, a truly immense area. At the beginning of the war it seemed to the Allies an impossible task to retake this conquered territory from the Japanese. And while the attack on Pearl Harbor on December 7, 1941, had badly damaged the U.S. Pacific surface fleet, it had not damaged the submarine fleet, which began offensive operations on December 8.

It took time for the U.S. Navy to recover from the attack, but with dogged determination and flexible, aggressive tactics, American subs slowly began to push back the far-flung outer ring of territories that had been captured and garrisoned by the Japanese. To survive, these garrisons required uninterrupted deliveries of food, weapons, and fuel in quantities that could be transported only by Japanese ships that were prime targets for U.S. submarine torpedoes. As ship sinkings mounted, the ring of territories with its fragile network of shipping lanes shrank until it collapsed.

As they had in World War II, Germany had waged all-out submarine war in World War I, both times taking a huge toll on Allied shipping. As in World War II, British ship losses in World War I had reached alarming proportions. The losses caused Admiral John Jellicoe, First Sea Lord of the Admiralty, to warn that the Germans would win the war unless the losses were stopped and stopped soon. Winston Churchill echoed Jellicoe's words twenty-five years later. Had the lessons learned by the British from the near disaster caused by the heavy loss of ships in both wars been heeded by the Japanese, the war in the Pacific might have lasted longer than it did. As it was the Japanese badly underestimated how hard it would be to maintain their lines of supply across the vast Pacific against U.S. submarines. To make matters worse, the Japanese had a weak convoy system and a weak antisubmarine force. Unlike the British and Americans in both wars, the creaky and inefficient Japanese convoy escort system could do little to protect ships that U.S. submarines were sinking faster than they could be replaced. By the time the Japanese got around to building an effective antisubmarine force the war had been lost.

Yet, in the war's early stages the U.S. submarine force found itself hobbled by an outmoded and conservative war-fighting doctrine that had been formulated during peacetime by submarine officers who had no combat experience whatsoever. It was hardly surprising, then, that this tactical relic from another era had to be jettisoned as soon as the bombs started falling on Pearl Harbor. Changes didn't happen overnight; it took time for the sub force to develop a new, aggressive doctrine based on tactics developed under actual combat conditions during war patrols. Once that happened, the Japanese merchant marine was doomed. The only thing that prevented it from being annihilated sooner than it was was the faulty torpedoes that plagued U.S. submarines at the start of the war and continued well into 1944.

The torpedo problem turned into a scandal that bordered on dereliction of duty by the officers in the Navy's Bureau of Ordnance (BuOrd). Their stubborn refusal to admit that the Navy's standard Mk 14 steam-powered submarine torpedo didn't always perform as designed—instead, blaming the inexperience of submarine crews for its problems—had a demoralizing effect on the force. The torpedo problem proved a tough nut to crack because the three separate faults inherent in the design of the Mk 14, working in concert, masked the fault each of them posed individually. It took over two years to isolate and fix the flaws residing in the Mk 14's depth-control device, its magnetic influence exploder, and its lightweight firing pin. The duds, erratic runs, and premature detonations of warheads caused by these different problems saved many a Japanese merchant ship from certain destruction and most certainly prolonged the war.

By early 1944, after extensive testing that included live torpedo shots and extensive modification and further testing of the faulty components, the sub force finally had a reliable weapon. Even so, problems continued to crop up even as the new and improved Mk 18 electric-powered torpedoes entered service. With better torpedoes, the sinking of Japanese ships increased dramatically, until, by early 1945, Japan's merchant marine had virtually disappeared from the Pacific along with its cargoes of rice, coal, iron ore, bauxite, rubber, and, the most important commodity of all, oil; in most Japanese cities automobiles had vanished from the streets, replaced by jinrikishas.

The Sea of Japan was the only area where what remained of the Japanese merchant marine blithely went about its business, unmolested by submarines. U.S. Pacific command understood that as long as Japan had ships to lug goods across that sea from occupied Manchuria and Korea to ports in western Japan, the war would continue, perhaps at a reduced tempo, but continue it would.

In 1943, Vice Admiral Charles A. Lockwood, Commander Submarine Force, United States Pacific Fleet (ComSubPac), had sent submarines on war patrols into the Sea of Japan. On three separate occasions they had slipped into and out of the sea by negotiating La Pérouse Strait, one of the five straits providing entry, and which the U.S. Navy believed was mined. The results of those operations were mixed, as torpedo problems plagued the five subs. Consequently, the USS *Permit* (SS-178), USS *Plunger* (SS-179), USS *Lapon* (SS-260), USS *Sawfish* (SS-276), and USS *Wahoo* (SS-238) sank only ten ships totaling approximately 28,000 tons, hardly enough to put a dent in Japan's *maru* lifeline. Worse yet, the *Wahoo*, on her second patrol into the sea, was attacked and sunk with all hands, including her famous skipper, Commander Dudley W. "Mush" Morton, revered by the sub force as the best of the best. With the *Wahoo*'s loss, Admiral Lockwood suspended further operations in the Sea of Japan.

And yet a different picture had slowly started to emerge in the spring of 1943, well before the first submarine patrols into the Sea of Japan. Scientists working for the Navy in laboratories in California had developed a radically new sonar system that held promise for locating mines underwater. Lockwood thought that with refinement this system might be useful to submarines for plotting the locations of mines in their patrol areas—including, perhaps, even the forbidden Sea of Japan, which, after the earlier penetrations, was now thought to be guarded by newer and even bigger minefields designed specifically to keep submarines out. Lockwood realized that the new system, called FM sonar (FMS), if fully developed, would give submarines a powerful offensive tool.

FM sonar evolved into that tool at about the same time hard-won victories in the Pacific suggested Japan might be nearing collapse. It seemed possible, then, that as ever more successful U.S. amphibious and surface naval operations gained momentum against Japan's army and navy, and while U.S. submarines with better-performing torpedoes sank more and more merchant ships, a timely raid by subs into the Sea of Japan to attack protected shipping might speed up Japan's collapse by cutting off supplies, which would demoralize her people and her leaders, who would then surrender.

ComSubPac assumed that such a raid would accomplish several things. First, it would demonstrate that by penetrating the minefields ringing the Sea of Japan, U.S. subs could operate virtually anywhere and under any conditions. Second, it would demonstrate that the Japanese were utterly defenseless against U.S. military forces. Third, besides cutting off vital war supplies, it would slow if not end the transfer of troops from Manchuria to Kyushu to meet the anticipated U.S. invasion of Japan. Fourth, it would demonstrate to the Soviet Union that the United States Navy's powerful submarine force would play a vital role in the implementation of America's strategic objectives in the postwar era.

As Admiral Lockwood surveyed developments in the sonar laboratories in California, he quickly grasped the tactical implications that the new device had unexpectedly presented. He wasted no time putting them to use. Thus were born the Hellcats and Operation Barney.

PART ONE

The Beginning of the End

CHAPTER ONE
A World Destroyed

The brick apartment building at 18 Collier Road, Atlanta, Georgia, still stands. Today it's in demand for its proximity to Midtown and Buckhead, and, across the street, a major medical facility, Piedmont Hospital. In July 1945 it was home to Sarah Simms Edge and her daughter, Sarah, age three.

With a second child due in August, Sarah, likely dreading another scorcher of a July day in Atlanta, went about her daily routine of household chores and caring for little Sarah. She was looking forward to the mail, anticipating a letter from her husband, U.S. Navy commander Lawrence Lott Edge. Since his deployment to the Pacific, Edge, commanding officer of the USS *Bonefish*, had written with such clocklike regularity that despite the great physical distance separating them, Sarah felt as close to her husband as was possible in wartime.

Sarah had last been with Lawrence in San Francisco, where the *Bonefish*, upon her return from the Pacific in early November 1944, had undergone a major overhaul at Hunters Point Naval Shipyard. Typically, Pacific Fleet subs underwent three- to four-month-long overhauls on the West Coast, either at Hunters Point or Mare Island Naval Shipyard. It had been a wonderful and long-anticipated reunion for the couple. Romantic, too, for Sarah soon discovered that she was pregnant.

After the *Bonefish* completed her overhaul in mid-February, she had sailed for Pearl Harbor. From Pearl Harbor she departed for Guam, arriving on Easter Sunday, April 1. Far to the northwest, in the Ryukyu Islands, American troops were landing on Okinawa, Japan's last Pacific bastion. Only 956 miles from Tokyo, the capture of Okinawa would give the United States a forward staging area from which to launch Operation Majestic, the proposed invasion of Japan, should it become necessary.

As the fight for Okinawa gained momentum, the *Bonefish* departed Guam on her seventh war patrol, Lawrence's third as CO. In addition to a regular combat patrol, he had orders to conduct a top-secret special mission. Well in advance of his departure from Pearl Harbor, Edge had been informed that this patrol would be his last as captain of the *Bonefish*, as he was slated for transfer to the staff of ComSubPac's new electronics training command. Edge had been trained in electronics and was looking forward to the challenges the job offered. Now, however, the letters he had mailed to Sarah before sailing from Guam hinted that upon completion of this next war patrol, yet another special mission was in the offing. Lawrence didn't try to hide his disappointment; he craved shore duty.

Beginning in early March, Lawrence's letters posted from Pearl Harbor arrived almost daily at Collier Road. They were filled with insightful and deeply personal reflections on his duties as the commanding officer of a submarine, and on the war and the effect its continuation was having on his and his crew's morale. Like the Atlanta newspapers and radio broadcasts reporting steady Allied advances against the Nazis and Japanese, Lawrence's views of the war in the Pacific gave Sarah good reason to believe that, yes, the war might be nearing an end, and that her husband would soon be home. But more than anything else he expressed deep love for Sarah and for his daughter, whom he sometimes called "Sarah, Jr." or "Boo." He yearned for the day he would come home to them.

His letters continued arriving from Guam before departure on his seventh war patrol and again after his return and on into late May as he prepared for the special mission hinted at in his earlier letters. Perhaps Sarah sensed that something was wrong when his letters, which had been arriving with clocklike regularity, suddenly stopped coming. Then, on that hot July day, she heard a knock on the door from a messenger with a telegram and a heart-stopping, "Ma'am, is someone with you?" The words that rose off the strips of teletype pasted on a yellow Western Union form cut like a knife.

WASHINGTON DC JUL 28 [1945]

MRS SARAH SIMMS EDGE ATLANTA

I DEEPLY REGRET TO INFORM YOU THAT YOUR
HUSBAND COMMANDER LAWRENCE LOTT EDGE USN
IS MISSING IN ACTION IN THE SERVICE OF HIS
COUNTRY. YOUR GREAT ANXIETY IS
APPRECIATED AND YOU WILL BE FURNISHED
DETAILS WHEN RECEIVED. TO PREVENT
POSSIBLE AID TO OUR ENEMIES AND TO
SAFEGUARD THE LIVES OF OTHER PERSONNEL
PLEASE DO NOT DIVULGE THE NAME OF THE
SHIP OR STATION OR DISCUSS PUBLICLY
THE FACT THAT HE IS MISSING.

VICE ADMIRAL RANDALL JACOBS CHIEF OF
NAVAL PERSONNEL.[1]

In an instant, that telegram turned Sarah Edge's world upside down
and changed her life forever. The Navy had no other information about
Lawrence beyond what was in the telegram, only those three terrible
words: *Missing in action.* Had the *Bonefish* been sunk? Was Lawrence
dead or alive? Had he been taken prisoner? She may have wanted to
scream, *Tell me!* But no one could, not even the Navy.

Like most Navy wives, especially those of submariners, Sarah Edge had
no idea, other than Pearl Harbor or Guam, where her husband was at any
given time. The little information she did have came from reading between the lines of his self-censored letters, and it wasn't much. He usually
started writing them aboard ship, continued them in his off-duty hours—
of which there were few for a sub's skipper—then finished and posted
them when he returned from patrol.

Lawrence had written one of the many letters Sarah received in April while the *Bonefish* was engaged in that first secret mission.

> *Lovely, dearest wife,*
>
> *. . . The patrol is moving along but slowly. . . . Actually there have been few dull days and too few dull moments, though I think we'll all be glad when this patrol is over. Luckily (and I haven't mentioned it to you before) it is to be a short one, because of a little special mission [we have been] assigned.*[2]

Unknown to Sarah, Lawrence at that time was conducting a dangerous reconnaissance of a minefield sown in waters near Kyushu, Japan. His mission was similar to those that had been assigned other sub skippers: to locate and plot enemy minefields using an experimental mine-detecting sonar system.

The experimental sonar gear in the *Bonefish* worked to near perfection, plotting strings of mines with impressive accuracy. After Lawrence's return to Guam from the mine recon he told Sarah that he and his crew were putting in long hours and working very hard, but that everything was okay.

> *My precious Darling,*
>
> *Today has been a long tiring one, and since we start early again in the morning, I'll only take time tonight to whisper again that I love you dearly and miss you intensely.*
>
> *. . . There is still no news to write about from out here. Time is drawing rapidly closer for departure on a new patrol, and practically all personal doings are in that connection.*

Lawrence didn't divulge the nature of his work or the mission he'd completed nor what it portended for the future, only that he would be starting another patrol soon, after which he hoped, at last, to transfer ashore. In late May, before he departed on this, his fourth patrol, he wrote a short letter filled with words of love for Sarah and Boo, and joy that he was the luckiest man in the world to have such a sweet and wonderful wife. He closed with:

> *Goodbye, my precious for today. You'll be constantly in my thoughts as well as my heart until I can write again. . . .*
>   *With all my deepest love,*
>   *Lawrence*[3]

———

Sarah had always feared that Lawrence would be killed in the war. Now that fear had materialized into a nightmare. The missing-in-action telegram had shocked her like a fall into icy water. Pregnant, racked by dread and anxiety, Sarah knew that submariners faced down danger and death every day. She may also have known that as of July 1945 the Navy had lost fifty submarines. Had the *Bonefish*, too, been sunk? In those first dark hours of uncertainty she may have recalled words that Lawrence wrote to her in April:

> *[T]here's the feeling that we've been lucky enough to survive so far; it would be such a shame not to last for the remainder and thus live through the whole thing.*[4]

All Sarah could do was wait in agony for the Navy to provide her and the families of the missing *Bonefish* sailors with more information about their fate. And pray for her husband and those men.

CHAPTER TWO
ComSubPac

Asubmariner himself, Charles A. Lockwood, short in stature and with a warm, outgoing personality, was a capable and respected career naval officer. His extravagant regard for the men under his command was repaid with their undying admiration and affection. To them he was "Uncle Charlie."

Not one to hide his feelings, Lockwood freely admitted that he was deeply affected by the deaths of the more than 3,500 officers and enlisted men serving in the fifty-two subs lost during the war.* To Lockwood, his submariners (that's sub-muh-REEN-ers, not sub-MARE-iners) were his family.

The force suffered its first loss on December 10, 1941, when Japanese planes attacked the navy yard, Cavite, Philippine Islands. The USS *Sealion* (SS-195), commanded by Lieutenant Commander Richard G. Voge (who would later serve as ComSubPac operations officer; his last name is pronounced *Vouge-E*), was undergoing repairs when it was hit by bombs and sunk. The attack killed five members of her crew. The last U.S. submarine lost in the war, the USS *Bullhead* (SS-332), commanded by Lieutenant Commander E. R. Holt, was bombed by a Japanese plane and sunk with all hands while patrolling in the Java Sea near Bali on August 6, 1945, the same day that the B-29 *Enola Gay* dropped an atomic bomb on Hiroshima. Lockwood, who had vowed that the Japanese would pay a heavy

---

*The U.S. Army Air Force from March through August 1945 lost approximately 3,000 combat air crewmen and 485 B-29 bombers in attacks on Japan.

price for every submariner killed, eventually kept his promise, which he regarded as an almost sacred obligation.

Charles Andrews Lockwood was born in Midland, Virginia, on May 6, 1890. Raised in Lamar, Missouri, Lockwood was a self-described country boy steeped in the lore of rural turn-of-the-century America. Though far from the sea, he knew even as a youngster that he wanted a career in the United States Navy. Appointed to the U.S. Naval Academy by Missouri senator William Stone, Lockwood entered the academy as a plebe with the class of 1912.

An average student, Midshipman Lockwood ranked academically in the lower half of his class. That ranking, plus his experiences at sea along the eastern seaboard of the United States during his class cruise aboard the old armored cruiser USS *Chicago* (CA-14), would, it seems, hardly have prepared him for a future career in submarines. Like most midshipmen of his day, Lockwood was determined to serve in a modern surface warship, preferably a battleship, certainly not a submarine, which, at that time, it was thought no self-respecting naval officer would want any part of. His first encounter with a sub, a primitive *Holland*-class boat, really not much more than a riveted sewer pipe of a vessel, left him shaking his head in disgust. Little did he know.

After graduation Ensign Lockwood served, first, in the old battleship USS *Mississippi* (BB-23), then in the newly commissioned USS *Arkansas* (BB-33). To Lockwood, a tour of duty aboard the new 562-foot-long, 26,000-ton-displacement *Arkansas*, with her twelve 12-inch guns in six turrets, personified the real spit-and-polish Navy. This assignment turned into a serious and eye-opening experience for the young officer, for when he was thrust into the *Arkansas*'s wardroom, filled as it was with high achievers, it put Lockwood's desire to succeed as a naval officer to the test. He passed the test, though not without having to overcome a few rough spots along the way, he said. He also said that this singular tour of duty had not only helped him mature, it had also formed his character and lifelong commitment to excellence.

As for the submarine service, Lockwood more or less stumbled into it. Transferred to Manila, Philippines, in 1914, he was talked into applying for submarine duty by a fellow officer who enticed him by pointing out that subs, with their less structured organization than that of the Navy's surface fleet, offered a speedier path to command at sea. Lockwood took the bait (every naval officer dreams of being addressed as "Captain" on the bridge of his own ship) and seemingly in no time at all had qualified for command of a so-called "pigboat," the gasoline-powered submarine *A-2*. Commissioned in 1903, the *A-2* was a cranky old tub in need of constant repair. And with its gasoline-fume-fouled atmosphere, it was a bomb waiting to go off. More than one sub had blown up or caught fire, with disastrous consequences for its crew. Yet for all of the problems Lockwood encountered as CO of the *A-2*, it was that old sub that set him on the path that, while it meandered from time to time, he would follow for the rest of his career.

In 1918, just as World War I drew to a close, Lockwood, who had seen no action in that war, received orders to Japan. While there he took note of the continuing emergence of Japanese imperialism, which had gained momentum with Japan's earlier acquisition of territory in Korea, and, during the First World War, the annexation of German-controlled Shantung, China. Further acquisition by Japan of the German Pacific island mandates of World War I emboldened Japan's leaders to pursue their unfettered imperialist aims in East Asia. Lockwood claimed to have had a premonition that in due time Japan would pose a serious danger to the United States that would one day lead to war between the two countries.

After completing his tour in Japan, Lockwood received orders to the Navy's sub base at New London, Connecticut. There he took command of another gasoline-powered sub, the old *G-1*. But not for long. Now a lieutenant commander, Lockwood transferred yet again, this time back to the Asiatic Fleet in Manila to command a river gunboat. After he completed this second stint in Manila, it was on to Rio de Janeiro as a member of the U.S. Naval Mission to Brazil. The interwar years of the late 1920s and early 1930s flew by in something of a blur, for with Lockwood's promotion to full commander, he arrived in San Diego to take command of Submarine Division 13. He was never in one place for very long, in 1937; a change of orders sent him to Washington, D.C., and the office of the Chief of Naval Operations to chair the submarine officers' conference, an organization that guided the development and design of new submarines.

Lockwood had become one of the Navy's top experts on submarines and submarine tactics. As chair he fought hard to change the narrow and hidebound thinking of senior officers who resisted many of the improvements the submarine force needed to undergo to modernize and prepare for the war with Germany and Japan that was then looming on the horizon. Lockwood, in collaboration with two equally farsighted colleagues, Lieutenant Commander Andrew I. McKee and Lieutenant Armand M. Morgan, tirelessly battled Navy red tape and bureaucratic intransigence to win approval for the development and construction of what would prove to be the supremely successful wartime *Gato*-class fleet-type subs. With only a few basic design modifications introduced during the course of their deployment, the *Gato*s and their sister *Balao*- and *Tench*-class boats became the workhorses of the Pacific submarine fleet.

In 1939 Lockwood received a promotion to captain and was made chief of staff to Commander, Submarine Force, U.S. Fleet, headquartered aboard the light cruiser USS *Richmond* (CL-9). Lockwood, who wanted sea duty above all else, once again had to endure a tour of shore duty, albeit an important one. His influence on the Navy's thinking about how to effectively build and utilize modern submarines in wartime had slowly, if not reluctantly, been adopted by a peacetime sub force undergoing the changes needed to prepare for war.

In February 1941, Lockwood, fresh from his duties as submarine force chief of staff, arrived in London as a naval attaché. He immediately set out to learn as much as he could about the war now raging in Europe (in London he got a taste of the Blitz), especially the war at sea where the Royal Navy was fighting a desperate, all-out battle against German U-boat commerce raiders in the North Atlantic. He was still in London when the Japanese attacked Pearl Harbor. All he could do was fume at being stuck on the beach out of the shooting war, while sending requests to the Navy's Bureau of Naval Personnel (BuPers) in Washington, virtually begging for an assignment at sea. His requests were denied because, said the bureau, it could find nothing for him to do, since almost all of the seagoing billets had already been filled, many of them, Lockwood noted sourly, by officers junior to him. Then things suddenly changed.

On March 5, 1942, Lockwood received a promotion to rear admiral and orders to proceed to Fremantle, Australia. There he was to relieve Captain John Wilkes, commander of the ragtag SubsAsiatic Force of the Asiatic Fleet that had been driven out of the Philippines by the Japanese. Lockwood was ecstatic. This was what he'd been craving all along: action in subs against America's main enemy. Not *in* subs exactly, but in command of subs, which was the next best thing, as Lockwood had never fired a torpedo in anger during a real war, just in exercises for war. On his arrival in Australia, Lockwood immediately assumed two hats: Commander, Submarines, Southwest Pacific Fleet (ComSubSoWesPac) and, temporarily, Commander, Task Force 51, a surface ship command. His arrival had precipitated other shifts in the evolving submarine command structure in Australia dictated by events in the Philippines and in Java, both of which soon fell under Japanese control. These changes included the reassignment of Rear Admiral Ralph W. Christie, in charge of submarines in Brisbane, to the Naval Torpedo Station at Newport, Rhode Island. Captain James Fife, head of the administrative staff of SubsAsiatic, replaced Christie in Brisbane, while an exhausted Wilkes returned to the United States for reassignment.

In early 1942 the submarine force, reacting to the requirements thrust upon it by the attack on Pearl Harbor and the Philippines, quickly reorganized from its original three commands into two. The SubsAsiatic force was decommissioned, putting Wilkes out of a job, and folded into the two remaining Pacific commands: ComSubSoWesPac, headquartered in Fremantle, Australia, and now headed by Lockwood; and Commander, Submarines, Pacific Fleet (ComSubPac), in Pearl Harbor, under the command of Rear Admiral Robert H. English. The two commands operating within their assigned areas of the Pacific would continue to function independently of each other but would share common tactical and technical attributes, including intelligence collection and distribution.

Thus, Pearl Harbor subs commanded by Admiral English (ComSubPac) operated in an area extending west from Pearl Harbor to the eastern coast of China, north to Hokkaido, Japan, and south to the Caroline Islands area. Fremantle subs (and for a short time those few subs still based in Brisbane with Captain Fife) commanded by Admiral Lockwood (ComSubSoWesPac) operated in an area that encompassed the Philippine Islands, the southern coast of China, all of Indochina, Borneo, Java, the Malay Barrier, and New Guinea in the Coral Sea area. Lockwood's SoWesPac area, with its far-flung conquered territories, had forced the Japanese to weave a tangled web of shipping routes. By necessity these routes were shorter and more compact than those in ComSubPac's area of operations. In either case these were immense areas in which to conduct war patrols, and submarines, whether departing from Pearl Harbor or Fremantle, had to endure long voyages to reach their assigned areas.

Despite his wearing two hats, Lockwood immediately focused his attention on reviving a dispirited force of submariners in Fremantle. Not only had they been driven south of the Malay Barrier by the advancing Japanese, but their main offensive weapon, the Mk 14 torpedo, had proven itself unreliable. Far too many Japanese ships were escaping from attacks by U.S. subs, which so far had little to show for their efforts. Even worse, the seemingly unstoppable Japanese army, after overrunning the Philippines, Malaya, and the Netherlands East Indies with their prize oil fields, seemed poised for a thrust south to Australia. Intelligence reports claimed that there were 200,000 Japanese troops in the Malay area alone. The only thing that stood between them and the Australians and New Zealanders were the open waters of the Indian Ocean. Lockwood's subs with their lousy torpedoes were in no shape to mount a robust defense. Fortunately, the threat to Australia never materialized. Lockwood and his submariners, breathing easier, concentrated their efforts on solving more immediate problems.

The introduction by Lockwood of the new sub force war-fighting doctrine and its implementation by a new breed of aggressive young skippers eager to fight the Japanese, coupled with Uncle Charlie's optimism and enthusiasm, began to have a positive effect. As noted earlier, the submarine doctrine in use at the beginning of the war was based on outmoded and conservative peacetime principles themselves founded on a strategy designed principally to sink an enemy's warships, not its cargo ships. In practice it proved useless under actual combat conditions in the Pacific, where the vagaries of geography and the wide dispersal of Japanese naval forces nullified the doctrine's effects. Moreover, its reliance on passive sonar for targeting, minimal periscope exposure, and an almost unshakable belief in the safety of deep submergence during daylight hours proved totally inadequate. Senior commanders at first failed to realize how essential it was to cut off the lines of supply Japan needed to sustain her garrisons in the Pacific, and that it wasn't the Japanese navy that had to be destroyed, but Japan's merchant marine.

Despite the meager sinkings of Japanese ships by U.S. subs early on, it didn't take long for those senior commanders and for the sub skippers on patrol to discover just how badly flawed their old war-fighting doctrine really was. With the experience gained from operating in enemy-controlled waters, submarine command shaped a new doctrine emphasizing innovative and daring tactics. These tactics were perfectly suited to the younger skippers fast replacing the older, conservative ones from the prewar era. New tactics emphasized attacks on merchant ships instead of warships; night surface torpedo attacks instead of submerged night attacks; the use of radar and sonar to track and attack targets; high-speed daylight surface patrolling to cover more territory; the use of more frequent and prolonged periscope observations; and much more.

As this new generation of skippers took command of the submarines coming off the builders ways at the Electric Boat Corporation in Groton, Connecticut; the navy yards in Portsmouth, New Hampshire, and Mare Island, California; the Manitowoc Shipbuilding Company in Manitowoc, Wisconsin; and later, Cramp Shipbuilding in Philadelphia, Pennsylvania, the tide began to turn against the Japanese. Even so, no one, least of all Lockwood, had any illusions but that it would be a long, hard road back and that a lot of mistakes would be made along the way.

As the new doctrine entered sub force planning and operations, Lockwood continued to make steady progress toward solving the Mk 14 torpedo dilemma and convincing the ossified and bullheaded BuOrd that the sub force had a serious problem with its main offensive weapon and that it had to be solved pronto.

Two major flaws in the Mk 14—a penchant to run deeper than set and a sometimes obstinate refusal to explode—had been traced in part to the weapon's faulty depth-control mechanism and its overly complicated magnetic influence exploder. Through extensive testing Lockwood and his technicians traced the problems to flaws in the design of both the depth controller and exploder. This evidence proved conclusively that the torpedo problem lay not with incompetent submarine fire-control personnel, as BuOrd had claimed, but with BuOrd itself. In time, and with more testing, Lockwood would make the further discovery that the poorly designed firing pin used in the exploder mechanism was too flimsy to withstand a collision between a torpedo warhead and a ship's hull, as it sometimes caused the pin to bend out of shape and jam in its guideway before it could contact the primer to set it off. It took a lot of time and hard work to remedy these three interrelated problems, which plagued sub crews on and off until the end of the war. Lockwood the doer found himself fully engaged in SoWesPac submarine operations when things suddenly changed again.

On January 21, 1943, at 6:50 a.m., the *Philippine Clipper*, a four-engine Martin M-130 flying boat on loan from Pan American World Airways to the Navy, approached the coast of California near San Francisco after a routine thirteen-hour flight from Pearl Harbor. Aboard were ComSubPac Admiral English, three of his senior staff officers, six other Navy passengers, including a nurse, and a civilian crew of nine.

The Pan Am base at Treasure Island in San Francisco Bay radioed the clipper to report heavy rain, fog, and high winds, and to advise that under such conditions it would not be possible for the plane to land before daylight and that it should divert to San Diego. At 7:15 the pilot radioed that they were on a course due west, back out over the Pacific. Seven minutes later he requested a navigation fix, after which the plane was not heard from again.

The clipper's disappearance remained a mystery until an air search team spotted its wreckage a week later in the Ukiah area near Boonville, California, ninety miles from San Francisco and twenty-two miles from the ocean. The big plane had come in low, shearing off treetops before it crashed into a mountain, killing all nineteen aboard. The bodies were found with the wreckage in a fire-blackened ravine. It took days and the cutting in of a road from the main highway through heavily forested terrain with bulldozers to remove the dead and to comb through the wreckage for any classified documents relating to submarine operations. The wreckage was later buried under tons of earth.[1]

English had flown to California to inspect submarine support facilities at Hunters Point and Mare Island, after which he had planned to inspect those at Dutch Harbor, Alaska, and at Panama. The shock caused by his death, and the deaths of his staff of experts on submarine engineering and weapons, swept through sub command. Lockwood in Australia learned of it in a newspaper report, not through official channels. It threw the entire command structure into chaos and left a big hole at SubPac in Pearl Harbor.

According to Lockwood, he had no desire to be named English's replacement. He immediately wrote a letter expressing his preference to remain in Fremantle, where he was closer to the submarine front lines than he would be in Pearl, and sent it to one of his former bosses, who was now chief of staff to Fleet Admiral Ernest J. King, Chief of Naval Operations (CNO) and Commander in Chief, United States Fleet (Cominch). The crafty Lockwood claimed that the job of ComSubPac, should it be forthcoming, would be a step backward for him. In truth, Lockwood coveted the position of ComSubPac and hoped to get it. When he was selected for the position by Admiral King, and after receiving orders on February 5 to pack his bags for Hawaii, he didn't hesitate for a moment. Lockwood apparently suspected that his selection had been partially influenced by old friends in high places in the Navy bureaucracy. But his friend, King's chief of staff, assured him, "You were selected on the platform that the officer best qualified to determine the submarine policy throughout the Pacific should be at Pearl Harbor."[2] That was true. To run the submarine war the Navy had not only picked the most qualified officer but also one of the best submarine experts in the world. Ralph Christie, recalled from Rhode Island, took over Lockwood's old job in Fremantle.

Lockwood arrived in Pearl Harbor on February 15, 1943, sporting on both sleeves the three gold stripes of a vice admiral. After settling in at the Makalapa Hill bachelor officers' quarters (BOQ) overlooking the sub base, Lockwood confronted a mountain of issues that included everything from personnel assignments, lack of spare parts, and the need for enlarged repair facilities, to torpedo shortages, especially of the new Mk 18 wakeless electric. There were also serious issues regarding intelligence collection and its interpretation and the need for better communications. The inspection of West Coast submarine support facilities, which had been postponed due to English's death and which Lockwood planned to complete in his stead, would have to wait until after he'd had a chance to familiarize himself with the late admiral's operations at Pearl. Busy as he was, Lockwood still found time to reevaluate overall submarine strategy and give consideration to refining it. There were problems, too, with the new tactical doctrine that needed immediate attention. Yet looking beyond these problems he saw that there existed certain opportunities that hadn't existed before, brought about by the influx of those young, eager, and aggressive skippers. Lockwood decided to put those skippers to the test. If they passed, as he believed they would, then there might soon be a way to hit the Japanese where they'd not been hit before.

CHAPTER THREE
The *Wahoo*'s Last Dive

As long as men have gone to sea, whether in peacetime or war, it is understood that ships will sink and sailors will die. It is a given that their remains—flesh and bones, wood and steel—will rest in the sea for eternity, because, as every sailor is taught, the sea never gives up its secrets, nor its dead.

Of course, that's no longer true, at least not since the discovery of the *Titanic* and other long-lost liners and warships. Today the question is: What other ships might yet be found and what mysteries might be solved by their discovery? The answer to part of that question arrived in 2006, 2007, and again in 2010 when the news media reported that five lost World War II–era U.S. Navy submarines had been located: The USS *Lagarto* (SS-371), the USS *Wahoo* (SS-238), the USS *Perch* (SS-176), the USS *Grunion* (SS-216), and the USS *Flier* (SS-250). Three of the five subs had been lost with all hands. Not only had the subs been found, they'd also been filmed by the dive teams who located them.

Of the five submarines, the USS *Wahoo* was arguably the most celebrated, even though another sub, the USS *Trout* (SS-202), was famous for having spirited from under the noses of the invading Japanese twenty tons of Philippine government gold bullion and silver coins for safekeeping at Fort Knox, Kentucky. (The plucky *Trout* was sunk on or about February 29, 1944, in action off the Ryukyus east of Formosa.)

As for the *Wahoo*, her commanding officer, Commander Dudley W. "Mush" Morton, a Naval Academy class of 1930 standout, personified the ideals of heroism American submarine skippers of his generation sought to emulate. He was uncommonly fearless and utterly tenacious in his pursuit of the enemy, at times even a bit reckless. Few if any of his peers could match his drive and nerve. From the day he arrived in the Pacific, Morton seemed determined to carve a name for himself in the annals of submarine history. According to submarine historian Theodore Roscoe, "If the philosophy of a combat submariner could be summed up in a single word, one would certainly suffice for Morton's: 'Attack!'"[1] To be sure, Morton possessed all of the traits a World War II submariner needed to succeed. And while the *Lagarto, Perch, Grunion,* and *Flier* were fighting subs with experienced skippers, they never had a chance to amass the combat record of Morton and the *Wahoo.*

In July 2007, a search team found the *Wahoo* in La Pérouse Strait, a body of water separating the upper tip of Japan's northern island of Hokkaido from the crab claw-like southern tip of the Russian island of Sakhalin (formerly Karafuto). The strait connects the Sea of Japan to the Sea of Okhotsk. The *Wahoo*'s wreckage lies in 213 feet of water close to where she was attacked and sunk by Japanese planes and ships on October 11, 1943. On September 20, in company with the USS *Sawfish* (SS-276), Morton had driven his submarine westward from the Sea of Okhotsk through La Pérouse Strait into the Sea of Japan in search of targets. He never returned.

Seven months prior to Morton's fateful September patrol, Charles Lockwood was still getting the feel of his new job as ComSubPac after the death of Admiral English. Along with English's duties, he inherited the admiral's staff, among whom was ComSubPac operations and intelligence officer Commander (soon Captain) Richard G. Voge, the former skipper of the USS *Sealion*, which, as recounted earlier, had been sunk at Cavite. Earlier, and with Lockwood's blessing, English had pulled Voge off his new command based in Australia, the former USS *Squalus* (SS-192) now renamed the *Sailfish,*[2] and installed him in Pearl Harbor.

Voge was an exceptional officer, brilliant, thoughtful, and articulate. He also had an uncanny ability, which he'd developed from close readings of intelligence reports, to anticipate Japanese merchant ship movements in time to redeploy patrolling submarines into position to attack them.

Lockwood had an extraordinary teammate in Voge, who served with him until the end of the war. As Lockwood's right-hand man, Voge had enormous influence on the evolution of submarine operations throughout the Pacific theater. In typical fashion, and because nothing escaped his purview, Voge had been keeping an eye on the all-but-landlocked Sea of Japan, with its shipping routes running arrow-straight between the Asian mainland and western Japan. Those routes had been inked in red on the big pull-down map of Japan on a wall in Lockwood's office. The Sea of Japan had yet to be exploited for targets by U.S. submarines, so it wasn't long before Lockwood and Voge began to plan a possible foray into its confined waters. The biggest challenge they faced in their planning was how to get submarines in there and how to get them out without being caught by the enemy.

The Sea of Japan is surrounded by mainland Asia, the Korean peninsula, and the Japanese islands. Its 250-mile width and 900-mile diagonal length covers an area of about 390,000 square miles. It has a maximum depth of over 12,000 feet, its bottom running up into rocky shallows against the western coast of Japan. In some respects Japan's coastline, with its small, rugged islands populated by seabirds and its deep inlets and big-shouldered bluffs, looks similar to the coast of Maine. Local weather conditions are influenced by the stormy Sea of Okhotsk and vary wildly by season and location. Gale-force winds, heavy fog, and below-freezing temperatures (La Pérouse Strait can freeze solid in the winter) rack the sea's mid and northern latitudes well into early summer, while in the south, milder temperatures and fair weather are more common.

Ships can enter the Sea of Japan from the East China Sea, the Pacific Ocean, and the Sea of Okhotsk through five straits that vary significantly in both width and depth. In the far north the Strait of Tartary lies wedged between the eastern coast of Siberian Russia and the western coast of Karafuto; the aforementioned La Pérouse Strait gives access to the Sea of Japan from the Sea of Okhotsk; the twisting, narrow Tsugaru Strait sits between southern Hokkaido and northern Honshu; the Shimonoseki Strait separates southern Honshu from northern Kyushu; the Tsushima

Strait, between the southern tip of Kyushu and the southeastern coast of Korea, is divided into eastern and western channels by the fortresslike Tsushima Island. Lockwood and Voge knew that the mission they had in mind would not be easy to execute nor free of risk. On the contrary, they knew that the natural and man-made obstacles lurking in the straits might be a death trap for submarines.

Voge had studied the intelligence reports supplied to ComSubPac by ICPOA (Intelligence Center, Pacific Ocean Areas), the Pacific Fleet's intelligence-collection operation based at Pearl Harbor. The reports convinced Voge that all five straits were patrolled by antisubmarine aircraft and patrol boats and sown with antisubmarine mines. Moreover, ICPOA believed that shore batteries lined both sides of the narrow Tsugaru and Shimonoseki straits. As for the Strait of Tartary, its northernmost approaches from the Sea of Okhotsk were often icebound a good part of the year. Russian destroyers patrolled the southern end of the strait.

Voge presented his findings to Lockwood. Declaring the Tsugaru, Shimonoseki, and Tartary straits unsuited for passage by submarines, Voge and his boss gave the Tsushima Strait and La Pérouse Strait a close look.

Because Tsushima was wider than the other four straits and had a deep-water trench, it seemed to be an ideal entry point for submarines. The problem with it was that ICPOA had little in the way of solid intelligence regarding the size and density of the minefields sown in the strait and its approaches. To further complicate matters the Tsushima Strait had a unique hydrographic feature known as the Kuroshio Current, the Japan Current, sometimes called the Black Stream for its deep blue color.

A branch of the equatorial current of the Pacific Ocean, the Kuroshio flows northeastward at a speed of approximately three knots along the coast of Formosa to Japan and thence into the Sea of Japan. The current continues northward until it reaches La Pérouse Strait, where it flows out into the Sea of Okhotsk. Voge pondered what hydrodynamic effect the inflowing or outflowing Kuroshio Current might have on a submerged submarine moving through the straits. Would it slow or speed up its passage?

The biggest drawback Voge saw to using the Tsushima Strait to enter the Sea of Japan was its minefields. Lockwood agreed. He believed that they posed a virtually impregnable barrier to penetration by submarines, surfaced or submerged, never mind the presence of antisubmarine air and surface patrols. Faced with having to penetrate an uncharted mine barrier, he and Voge focused instead on La Pérouse Strait. Intelligence reports indicated that it had a safe, unmined channel that the Japanese allowed neutral Soviet surface ships to use when sailing between the Sea of Okhotsk and the Russian naval base at Vladivostok on the Siberian coast. Intelligence indicated that the Japanese had sown mines in the La Pérouse channel at various depths between forty feet and seventy feet. This allowed Russian merchant ships—Japanese, too—to make safe passage, as very few ships in the world other than warships drew more than thirty-five feet of water. The mines, then, posed a danger only to submerged submarines, not surfaced ones.

Intelligence regarding the La Pérouse mine plants had been gleaned from various sources, including Japanese sailors picked up at sea from ships that had been sunk by subs operating close to Japan; from notices to mariners in the form of published bulletins distributed by the Japanese to local shippers, warning them of mined waters to be avoided; and from decrypted enemy radio broadcasts containing similar information transmitted to Japanese ships in the vicinity of the Sea of Japan. ICPOA had also picked up a few sketchy reports from spies who had made contact with Russian seamen familiar with La Pérouse Strait. Lockwood reasoned that if Russian ships could transit the strait via the safe channel, then so might surfaced American submarines. He also reasoned that even if a sub skipper didn't know exactly where the channel was he could always trail a Russian ship whose master knew what course to steer.

After sifting through these reports and weighing the pros and cons, Lockwood and Voge came to the same conclusion: It was high time that U.S. subs poked their bows into the Sea of Japan.

As Lockwood's plan to get subs into the Sea of Japan began to jell, he took time out to undertake the West Coast inspection tour Admiral English's death had left unaccomplished. From Pearl Harbor Lockwood flew to Dutch Harbor, Alaska, then to San Francisco, ending his tour in San

Diego. The submarine-repair facilities he toured were not as well equipped nor as efficient as Lockwood had hoped for. A lack of facilities and especially of trained civilian personnel hampered efforts to repair submarines and return them to the combat zone as fast as possible. Lockwood realized that it would take a special task force dedicated to the job of revamping the facilities and speeding up the workflow to accomplish what was needed.

While in San Diego Lockwood received an invitation from Dr. Gaylord P. Harnwell, director of the University of California Division of War Research (UCDWR), to visit the facilities Harnwell directed at the U.S. Navy Radio and Sound Laboratory (USNRSL) at Point Loma. UCDWR had been established in early 1941 to carry out research and development for the Navy under a contract let by the National Defense Research Committee (NDRC). Harnwell was a respected physicist and educator from Cambridge and Princeton universities. He had also taught physics at the California Institute of Technology and, before the war, had headed the physics department at the University of Pennsylvania. Steeped in advanced electronics and engineering concepts, he wanted to demonstrate for Lockwood some of the new prosubmarine gear, as it was called, undergoing development at UCDWR's labs. Because prosubmarine gear was high on Lockwood's list of priorities the sub force needed to counter Japanese antisubmarine measures, he eagerly accepted Harnwell's invitation to see what progress had been made in that area.

Earlier in the war Harnwell's efforts had been focused primarily on the problems associated with what was then called "subsurface warfare," a discipline that encompassed not just antisubmarine weaponry but also mine detection. As the tempo of Pacific submarine operations increased, the need arose for devices to enhance the inherent stealth of U.S. submarines and to improve their survivability in combat. Harnwell and his staff of scientists had been working on such devices and were eager to demonstrate them for Lockwood's benefit. Not only did they have a rapt audience but also one who understood how difficult such specialized work was and how much time it had taken to develop the prototypes of the gear on display.

Harnwell demonstrated just about everything in UCDWR's inventory.

It included a device that could detect an incoming torpedo's range and bearing; a sonar decoy similar to that of the German U-boat "pil-lenwerfer," a cartridge filled with calcium hydride that when mixed with seawater produced huge quantities of hydrogen gas that bubbled like an out-of-control Alka-Seltzer tablet to confuse enemy sonar; ultrasensitive long-range passive—listening only—sonar for the detection of enemy ships; a bathythermograph, a device that recorded the varying temperature layers in seawater, which, because it had a masking effect on sonar, made submarine detection more difficult than it already was; depth-charge direction and range indicators; and, to help reduce a submarine's self-generated noise signature, a propeller cavitation warning device. Amazed at what he saw, Lockwood lamented that the sub force still lacked these "Alice in Wonderland" gadgets, which he believed would have vastly increased sinkings of Japanese ships and no doubt have saved hundreds of American lives. All he could do was put the devices on his wish list; he knew that despite the best efforts of Harnwell's scientists, it would take many more months of testing and refinement before the advanced prosubmarine gear reached the submarine fleet—if it ever did.

After conducting a tour of the lab, Harnwell sprang a surprise on Lockwood. Harnwell and his assistant, Dr. Malcolm Henderson, who, like Harnwell, was a brilliant, dedicated physicist, took their guest for a cruise off Point Loma in San Diego Harbor to demonstrate an experimental sonar device capable of detecting underwater mines and which was slated for installation in navy minesweepers. Henderson had led the group of scientists responsible for creating the device and was eager to demonstrate its abilities. Despite being a prototype with a spaghettilike breadboard electrical layout, the thing performed well enough to give Lockwood a peek into the future, even though, he admitted later, he failed to grasp its significance.

The device Harnwell had demonstrated for Lockwood was called FAM-PAS, or Frequency and Mechanically Plotted Area Scan. Its name would soon be changed to Frequency Modulated Sonar and later shortened to FM sonar, then just FMS. The device used a then-new audio technique designed for commercial radio broadcasting called frequency modulation. Unlike AM, or amplitude modulation broadcasts, in which the audio carrier wave frequency is constant, the FM audio carrier wave varies in frequency. Conventional sonar units of the 1940s used a short pulse of sound transmitted on a constant frequency followed by a long silence as the return echo from a detected object was converted into an audible sound on the same frequency. By contrast, FM sonar emitted a steady, continuous signal modulated to avoid interference between the outgoing signal and the returning echo from a detected object, thus ending the requirement for a time lag. The experimental unit Harnwell demonstrated could locate submerged objects of every description, including shoals, sandbars, kelp, steel submarine nets, and even steel-hulled ships. In an earlier test it had even detected dummy mines. Realizing that such a device would interest Lockwood, Dr. Henderson explained that though FM sonar was originally developed for use by minesweepers, it could also be used by submerged submarines to plot a course into a defended enemy harbor.

Though Lockwood was impressed with Harnwell's and Henderson's Alice in Wonderland invention, he didn't see how the device could be used by a submarine. At that time submarine targets were concentrated in deep-ocean areas of the Pacific where an attacking sub had room to maneuver and to evade Japanese escort vessels. The harbors where Japanese ships sometimes took refuge from attack were in most cases too shallow for submarines to enter without being detected, even if they had a device that could plot a course into and out of a harbor, so it wouldn't make any difference. For Lockwood, FM sonar was an interesting gadget that not only proved the scientists at UCDWR were inventive, but also that the Navy's substantial financial investment in the lab was starting to pay off.

Lockwood returned to Pearl Harbor buoyed by what he saw at UCDWR but disappointed by the long gestation period needed to develop the devices and to manufacture them in quantity for use by the sub force. Nevertheless, he reckoned that with his visit to UCDWR the force had established one of its most valuable contacts with the scientific world, and with Harnwell and Henderson, who would later play major roles in future submarine operations. After writing a report on his trip for distribution to Admiral King, Admiral Nimitz, and his own immediate staff, Lockwood put aside what he'd learned in San Diego for more pressing matters. At the top of his list was the get-acquainted look-see patrol of the Sea of Japan by a task force of submarines.

Lockwood wasn't sure what to expect from such a mission. He knew that Japanese ships were plying their routes in the Sea of Japan, carrying essential cargoes of food and raw materials back to mainland Japan. The big question was, How many ships were actively involved in this work, given that the bulk of the empire's merchant marine was busy elsewhere in the Pacific? How plentiful would targets be in the Sea of Japan? Would there be enough to justify the risks entailed in sending a couple of subs up there on a raid? He and Voge wouldn't know until they tried it. Furthermore, if submarines suddenly showed up in the emperor's sea, would the Japanese rush to block its exits, trapping the subs inside until they ran out of food and fuel, to be hunted down and sunk? Lockwood mulled these questions over for a time, then decided that the risk was worth taking, if for no other reason than it would provide the Navy's high command with vital information on the state of Japanese resupply operations at home. That information might influence future planning for an invasion of Japan, which would likely be necessary to end the war.

To avoid a prolonged operation in the Sea of Japan that would give the Japanese time to mount an aggressive defense, the plan Lockwood and Voge devised called for only a four-day hit-and-run raid that might just catch the Japanese napping and that would end before they could rouse their antisubmarine forces in strength. In May, Lockwood submitted his plan to Admiral King and Admiral Chester W. Nimitz, Commander in Chief, United States Pacific Fleet (CinCPac), for approval. A week later it came back with just one word over the admirals' signatures: "Approved."

In early July 1943, three submarines departed from Pearl Harbor bound for what some observers believed was a suicide mission into virtually unknown territory. Lockwood didn't agree, though the uncertainty of it all stoked his three-pack-a-day habit. Two of the three subs were the older prewar-built USS *Plunger* (SS-179), commanded by Lieutenant Commander Raymond H. Bass, and the USS *Permit* (SS-178), commanded by Commander Wreford G. "Moon" Chapple. The third sub was the new-construction USS *Lapon* (SS-260), making her first war patrol under Lieutenant Commander Oliver G. Kirk. Bass and Chapple were seasoned skippers with exemplary combat records; Kirk, also a veteran, had less combat experience than either Bass or Chapple. Another submarine, the big, old prewar USS *Narwhal* (SS-167), skippered by Commander Frank D. Latta, was dispatched to create a diversion by bombarding the island of Matsuwa To in the Kurile Islands northeast of La Pérouse Strait. Shells lobbed ashore from her twin six-inch guns would keep the Japanese busy and help the three subs make their getaway after completing their mission.

The Sea of Okhotsk, even in summer, is cold and fogbound. The subs groped their way west through the Kuriles for La Pérouse Strait, where they made their run-in on the surface at night at full speed, dodging fishing craft and navigating the safe channel by sheer guesswork. Most but not all of the Russian ships they saw had their running lights on to identify themselves as neutrals. Since none of the subs hit a mine, they and Lockwood had evidently guessed right about the location of the safe channel and the depth of sown mines. Bass, Chapple, and Kirk agreed that it was a hair-raising trip.

The submarines took up their assigned areas and at the appointed hour (Chapple in the *Permit* jumped the gun) began looking for ships to torpedo. As it turned out bad weather and intermittent problems with the raiders' vital SJ radar spoiled any chances they may have had to wreak havoc on the Japanese. Not only that, targets worthy of torpedoes proved scarce: Lockwood's suspicion that the bulk of the Japanese merchant marine was busy in the greater Pacific turned out to be correct. The *Permit* and the *Plunger* sank only three ships totaling roughly five thousand tons; the *Lapon*, already bedeviled by SJ radar problems, suffered from an inoperative Fathometer, which impeded skipper Kirk's determination to hunt for targets in foggy coastal waters off the coast of Korea. As it was, the *Lapon* encountered only sampans. Complicating matters, the *Permit* shot up a Russian trawler by mistake off Karafuto. Chapple, realizing his error, pulled thirteen survivors, including five women, from the frigid water. He considered landing them on Russian Kamchatka, but after a flurry of radio messages to Pearl Harbor, Chapple was ordered instead to off-load his passengers in Dutch Harbor, Alaska, to avoid a direct confrontation with angry Soviet authorities.

All in all, the first foray by U.S. subs into the Sea of Japan didn't break the empire's back. Lockwood's endorsement to the *Lapon*'s patrol report said it best: "Results were disappointing. 22 days of foggy weather did much to render operations difficult. . . . Contacts were meager."[3]

As for the *Narwhal*, delayed by lousy weather, she arrived off Matsuwa To long after the three raiding subs had departed for home. Nevertheless, Latta swung into action. After dodging a *Wakataki*-class destroyer and a Mitsubishi 96 bomber, the *Narwhal* encountered a spate of clear weather and took advantage of it to battle surface six miles off the beach, and at

2020 . . . began bombardment of air field on Matsuwa To. Red rays of setting sun still visible in northwest. Eastern edge of Matsuwa in shadow. NARWHAL against dark background of eastern sky . . . Our fire was slow but seemed accurate, using hangars as a point of aim. . . . Fire was concentrated on hangars and landing strip, hoping to keep planes grounded. . . . One large fire started ashore. Enemy returned fire after about four minutes from one gun and in ten minutes, fire was being returned from a battery of about seven five-inch guns. . . . Return fire was haphazard.

2030 Enemy return fire getting closer in range, shells whistling overhead.

2031 Enemy in range, splashes dead ahead . . .

2032 Secured guns, changed course away and dived.[4]

"Where we had expected our submarines to find an abundance of targets, they found few worth the expenditure of a $10,000 torpedo," said Lockwood. Disappointed with the results, he was cheered that the four subs had returned unscathed.

He and Voge immediately began planning a second mission, one that would build on and take advantage of the lessons learned. The most important lesson was that U.S. subs could operate in the Sea of Japan, proving that no part of the empire, no matter how remote, was immune from attack. Lockwood believed that the next time subs entered the sea they would sink everything in sight. He might have kept his fingers crossed that when they did, the problems still plaguing Mk 14 torpedoes would be solved once and for all. If the submariners were going to risk their necks again getting into the Sea of Japan, they had better have a reliable weapon to take along.

Ray Bass, skipper of the *Plunger*, immediately volunteered to go back in. Another volunteer was Mush Morton, skipper of the *Wahoo*, which had just arrived in Pearl Harbor after a West Coast overhaul. Even then Morton was one of Lockwood's stars. He'd already sunk sixteen *marus* totaling about 49,000 tons and was clearly on his way to outpacing every skipper in the force.

Lockwood gave Bass and Morton the okay and saw them off from Pearl Harbor on August 8. Lockwood had confidence that the experience Bass had gained from the earlier raid would come in handy. According to their operation orders the *Wahoo* and *Plunger* were to enter the Sea of Japan through La Pérouse Strait on August 14 at night and on the surface via the safe channel that the *Plunger*, *Permit*, and *Lapon* had used in July and had plotted so other subs could use it in the future. According to the latest intelligence assembled by ICPOA, there was no reason to believe that the Japanese had made any alterations to the layout of the existing minefields or that they had sown any more mines below the surface of La Pérouse Strait. Like the *Plunger*, *Permit*, and *Lapon* that had ventured in before them, as long as the *Wahoo* and *Plunger* didn't submerge in the strait and didn't stray from the confines of the channel, Bass and Morton had nothing to fear. Furthermore, if ICPOA developed any information that the Japanese had altered the minefields, SubPac would alert the skippers by radio.

The *Plunger* **developed engine and** motor problems that crippled her port propeller shaft, forcing her to steam on only one screw, which delayed her entry into La Pérouse Strait for two days. Rather than turn back, Bass decided to run the strait submerged during daylight rather than on the surface at night, which would require speed and maneuverability that the *Plunger* lacked. It was a death-defying decision on Bass's part, given that any mines anchored at seventy feet—never mind any moored at forty feet—allowed only a five-foot margin of safety between the tops of the mines and the *Plunger's* keel, which, when she submerged to periscope depth, measured sixty-five feet below the surface. It's not clear whether Bass, knowing the high-low layout of the minefields, simply threw caution to the wind or if he assumed that the force exerted on the anchored mines by the outflowing Kuroshio Current, which made them lean over (a phenomenon called "mine dip"), gave him an extra margin of safety. Whichever it was, somehow the *Plunger* with her frozen port shaft got through in one piece.

Unlike the first raid, this time targets were plentiful. Regrettably, torpedo performance was abominable. Bass, plagued with faulty Mk 14s, managed to sink only two cargo ships. Morton, meanwhile, was bedeviled with duds, broachers, and erratic runs. His and Bass's patrol reports are a litany of miss, miss, dud, broach, miss, miss. . . . Furious, Morton sought permission to end the mission and return to Pearl Harbor to have his remaining torpedoes examined. Lockwood, deeply disturbed by this unexpected turn of events, approved. So far the two forays had been miserable failures.

Back in Pearl Harbor, Morton barely contained his anger when he powwowed with Lockwood to curse the lousy torpedoes he'd lugged all the way to the Sea of Japan. His boss was sympathetic—he'd been wrestling with torpedo problems ever since taking over in Fremantle, and more than anyone, Lockwood wanted Morton to succeed. When the irate skipper insisted on going back with a load of freshly overhauled Mk 14s and a batch of new Mk 18 electrics, Lockwood gave his approval. This time he tapped the USS *Sawfish* (SS-276), skippered by Lieutenant Commander Eugene T. Sands, to accompany the *Wahoo*. The *Sawfish*, like the *Wahoo*, carried a mixed load of Mk 14s and Mk 18s.

Both submarines departed Pearl Harbor on September 10, arriving off La Pérouse via different routes. The *Wahoo* headed in first, then made tracks to patrol her assigned area in the southern part of the sea; the *Sawfish*, assigned the northern part, followed three days later. Morton had orders to conduct operations as he saw fit, after which he was to depart the area on October 21 and report his position by radio after transiting the Kuriles. The *Sawfish*, which during the patrol had no contact at all with the *Wahoo*, kept to her own schedule. Lockwood and Voge expected that the Japanese would be on guard for more submarine intruders and had so briefed Morton and Sands.

There were plenty of targets, but once again poor torpedo performance thwarted success, at least for the *Sawfish*. Sands made numerous attacks that should have resulted in sinkings. Instead he was plagued by duds and erratic torpedo behavior. He reported that on firing, several of his Mk 18s struck the bow torpedo tube shutters—the streamlined outer doors covering the tubes' muzzles—sending them wildly off course. Fortunately the tin fish were designed to arm only after completing a four-hundred-yard run or they might have blown the *Sawfish* to bits. In seven attacks on eighteen ships Sands failed to sink a single one.

On October 9, after sixteen miserable days on station, Sands, with Lockwood's permission, pulled out with several Mk 18s still in their tubes for the torpedo experts to examine. Making good speed through La Pérouse, the *Sawfish* encountered patrol boats and twice escaped a bombing by patrolling aircraft.[5]

The *Wahoo*, meanwhile, had vanished.

On October 5, four days before the *Sawfish* departed enemy waters, the Japanese news agency Domei announced the torpedoing of the eight-thousand-ton passenger steamer *Konron Maru* in the Tsushima Strait.

Lockwood acknowledged that since there were no other subs in that area it had to have been the work of the *Wahoo*. Unlike the *Sawfish*, maybe this time her torpedoes had worked.

On October 7, the *Honolulu Star-Bulletin* published an article about the attack. It read in part:

> An Allied submarine, slipping boldly into the waters off Japan's west coast, sunk a Japanese steamer Tuesday in an attack which took the lives of more than 500 persons, Tokyo broadcasts said today.
>
> There was little doubt the submarine was American. . . .
>
> Despite the strenuous efforts by warships and naval planes to rescue passengers and crew, Tokyo said only 72 of 616 persons aboard have been reported saved.
>
> Rough seas and communication trouble were said to have hampered rescue work. The announcement said the steamer was hit by a single torpedo and sank "after several seconds.". . .[6]

When the *Wahoo* didn't report to Pearl Harbor by radio as scheduled on October 21, Lockwood began to worry. Of course, she could have been delayed as the *Plunger* was by an engineering casualty that had forced the *Wahoo* to lie doggo for a few days while her crew made repairs. Lockwood chain-smoked and waited. As the days dragged by without a word from Morton, Lockwood feared that something serious had happened to the *Wahoo*, something he didn't want to admit was possible, even as he held out hope that she might limp into port, perhaps damaged but still intact. After all, he knew from experience that whenever the Japanese thought they had sunk an American sub—and they thought they had sunk hundreds—they were quick to announce it, even though more often than not the submarine had returned to its base with all hands safe aboard. In an unusual step, Admiral Nimitz approved a request by Lockwood for a search by air of the most likely route home the *Wahoo* would travel. A careful search of an area hundreds of miles west of the Hawaiian Islands returned without finding a trace of the sub. Distraught, Lockwood confided to his diary, "No news of Mush. This is the worst blow we've had and I'm heartbroken. God punish the Japs! They shall pay for this. . . ."[7]

In early November Lockwood had no choice but to report that the *Wahoo* was overdue and presumed lost with all hands. Reluctantly he took down the little magnetic submarine silhouette engraved with the *Wahoo*'s name from the wall map in his office. He had a hard time accepting that Dudley Morton and his crew were gone and it left him badly shaken. For Lockwood, Morton's and the other skippers' raids into the Sea of Japan, or Lake Hirohito, as it was now dubbed by U.S. submariners, ranked up there with German U-boat ace Günther Prien's raid on the British fleet at Scapa Flow in 1939 and with Jimmy Doolittle's raid on Tokyo in 1942.

With little information to go on and lacking U.S.-decrypted radio messages from Japanese antisubmarine forces, Lockwood naturally blamed the *Wahoo*'s loss on mines. He was convinced that the Japanese, stung by the two earlier forays by U.S. subs, and perhaps anticipating future incursions, had sown more contact mines in the straits, one of which must have destroyed the *Wahoo*. Unknown to Lockwood, Morton had sunk four ships, among them the aforementioned *Konron Maru*. His bold attacks had indeed put the Japanese on full antisubmarine alert. Yet the alert had not, as Lockwood assumed, prompted more mine laying.

The mystery of the *Wahoo*'s loss was solved at the end of the war, when intelligence officers combing through tons of Japanese documents seized at the naval ministry in Tokyo for information on the fate of missing men and ships found a report describing an attack on a submarine in La Pérouse Strait. The date of the attack matched the date that the *Wahoo* was supposed to make her exit.

The report stated that on October 11, at 0920, a floatplane had spotted an oil slick on the surface of the strait. The *Wahoo*, perhaps damaged by an earlier depth charging or a bomb, must have been leaking diesel fuel. Then at

0945 Circling, the pilot identified a black conning tower and after calling in more planes, dropped a bomb on what he described as a black hull with a white wake. A second bomb brought up more oil. Aircraft number two arrived and dropped four small bombs which brought up more oil.

1025 Second floatplane dropped more bombs.

1135 A floatplane guided Submarine Chaser No. 15 to the area of the attack. It dropped nine depth charges followed by seven more.

1207 In the eruptions a large piece of bright metal identified as a propeller blade was seen.

1221 Another submarine chaser arrived and dropped six depth charges.

1350 Searching aircraft reported that neither the submarine nor her wake was visible.[8]

The *Wahoo* had made her last dive.

After the loss of the *Wahoo* Lockwood grappled with several important issues that would have an influence on decisions he would make regarding future raids into the Sea of Japan. Despite the small number of ships sunk on the three raids launched so far (he nevertheless regarded every sinking as another nail in the emperor's coffin), he had to question whether the experience the sub force had gained toward future raids had been worth the lives of Morton and his crew. It might have been hard for Lockwood to weigh this objectively, for despite his sorrow at the loss of so many other submarines and their crews and his deep affection for every man in his force, Morton was clearly Lockwood's favorite. He was unique.

He possessed all the skill, daring, and tenacity that Lockwood sought in his skippers and that Morton had amply demonstrated during the battles he'd waged against the Japanese on earlier war patrols.

Lockwood expressed his deepest feelings about Morton when he said, ". . . I resolved, there would come another day—a day of visitation—an hour of revenge. In time we would collect for the *Wahoo* and Commander Dudley Morton and his men, with heavy interest. And in time we did."[9] He had no way of knowing whether there had been any survivors from the *Wahoo*, but experience told him that it was unlikely. When submarines were hit, they went down fast—too fast for the men inside to escape. With the *Wahoo*'s loss, Lockwood reluctantly decided to put an end to operations in the Sea of Japan—for now, at least. He knew that if his subs were to return to sink ships and exact revenge for Morton and the *Wahoo*, they would have to have special equipment that could accurately plot the location of the minefields to give the raiding submarines a greater margin of safety. Lockwood knew that a submarine's greatest attribute is stealth. He believed that if submarines could find a way to penetrate the minefields submerged without alerting the enemy to their presence, next time they would deliver a blow from which the Japanese might not recover.

As the pain of Morton's death slowly eased, Lockwood thought about his visit to the UCDWR labs earlier in the year and the demonstration of FM sonar that had been conducted for his benefit off Point Loma. In his mind a bell began to ring, softly at first, then louder, until its peal began to slowly but surely unleash a chain of events that would culminate in what would come to be known as Operation Barney. It also unleashed a chain of events that would have a profound and lasting effect on the lives of a select group of submariners and, most especially, those of the USS *Bonefish*.

CHAPTER FOUR

The Commander from Georgia

On March 7, 1943, Mrs. Freeland A. Daubin, wife of the former Commander Submarines Atlantic (ComSubLant), smashed a bottle of champagne on the bow of the USS *Bonefish*. Duly christened, the submarine slid down the building ways of the Electric Boat Corporation in Groton, Connecticut, on a sunny but cold day, pennants and red-white-and-blue bunting flapping in the wind as she entered the Thames River stern-first. She was one of those good-looking *Gato*-class fleet-type submarines: a long, beautifully proportioned vessel, her low freeboard sweeping back over a 312-foot length and 27-foot beam to a narrow and trim stern. With her sharply raked snout and rounded gray body shaded with black, the *Bonefish* in many ways resembled the Florida game fish after which she was named.

Sitting in the water fully loaded with fuel, ordnance, and food, the *Bonefish* displaced 1,800 tons and drew 17 feet of water. Divided into nine watertight compartments, they included two engine rooms housing four General Motors Winton 16-248A 16-cylinder, 1,600-horsepower diesel engines. Each engine drove a DC generator (thus, diesel-electric drive) that supplied electricity to the ship's four main propulsion motors and for lighting, air-conditioning, radar, and communications, among other things.

Connected through reduction gears to the ship's twin propeller shafts, the motors drove the *Bonefish* at a top speed of twenty-one knots on the surface. Submerged, two massive 126-cell rechargeable electric storage batteries provided sufficient power to propel the *Bonefish* at nine knots, though for only a short duration and with the penalty of a high discharge rate. A more usable underwater speed of two or three knots, though limiting maneuverability, conserved electricity and provided greater submerged cruising range. When the submarine returned to the surface one or more diesels recharged the batteries while those not used for battery charging drove the ship. When all four main engines went online for high-speed surface cruising the small auxiliary diesel picked up the extra electrical load needed to keep the batteries topped off.

The *Bonefish*'s main armament consisted of ten 21-inch torpedo tubes: six forward, four aft. She carried ten war shots in her tubes and fourteen reloads. A pair of 20mm guns augmented the old-style long-barreled four-inch .50-caliber gun mounted on her forward deck. Later, from experience gained during war patrols, submarines employed heavier armament, usually a five-inch .25-caliber deck gun and twin 40mms. Submarine command didn't encourage the use of deck guns because return fire from enemy ships could cause serious damage to the sub's pressure hull. Also, because of the sub's low freeboard, exposed gun crews ran the risk of being swept overboard, if not killed or wounded by enemy fire. Yet there would come a time late in the war when Japanese ships were so scarce that submarines assumed the role of submersible gunboats, damaging and sinking Japanese small craft not worth the expenditure of a torpedo.

Big submarines like the *Bonefish*, with their usual complements of eighty-one men—seven officers and seventy-four enlisted—were well designed for their offensive role in the Pacific, given that many of Admiral Lockwood's ideas had been incorporated into them. With a fuel capacity of almost 90,000 gallons and a cruising range of 13,000 miles, a fleet-type submarine could easily undertake a sixty- to seventy-day war patrol to hunt down and sink Japanese ships.

After fitting out and undergoing sea trials the *Bonefish* was commissioned on May 31, 1943, under the command of Lieutenant Commander Thomas W. Hogan. During four war patrols out of Fremantle, Australia, Hogan, an aggressive and resourceful skipper, sank seven ships* for a total

---

*Of the seven ships sunk by Hogan, the 4,645-ton *Suez Maru* had several hundred sick Japanese soldiers and hundreds of sick British and Dutch POWs aboard, many of them stretcher cases. The *Bonefish* torpedoed her off Surabaya, Java, on November 29, 1943.

of 34,329 tons.[1] During that period the *Bonefish* compiled an enviable record that included Navy Unit Commendations for her first, third, and fourth war patrols. Little more than a year later, on June 13, 1944, in a ceremony attended by ComSubSoWesPac Vice Admiral Ralph W. Christie, Lockwood's replacement, Hogan relinquished command of the veteran *Bonefish* to thirty-two-year-old Lieutenant Commander Lawrence Lott Edge, USN.

Lawrence Edge had graduated from the United States Naval Academy in 1935. Tall and slender, he had a prominent forehead topped by thinning dark hair (his portrait in the *Lucky Bag* class yearbook foreshadowed this early baldness). The *Lucky Bag* remarked on Edge's pleasant personality, soft Southern drawl, and deep inner nature. The latter might account for his musical artistry, both on piano and violin, his interest in photography, woodworking, and drawing and painting. His watercolors of tropical flowers are sublimely beautiful. One might think Edge would have been more at home on the faculty of a small college, happily ensconced in an office strewn with books, than in the hot conning tower of a submarine filled with sweating sailors. Not so.

Upon graduation from the Naval Academy, Ensign Edge served a two-and-a-half-year tour of duty in the battleship USS *Maryland* (BB-46). It may have been that the "real Navy" of battleships and cruisers that so appealed to Charles Lockwood appealed less to Edge's deep inner nature than one might expect inasmuch as he requested duty in submarines, which even then was a service still shunned by officers with ambitions to flag rank.

After selection for duty in subs Lawrence reported to the U.S. Navy's modernized and bustling submarine school at New London, Connecticut, in January 1938. A week after graduating from sub school, on June 15, 1938, he and Sarah Simms married in Atlanta, Georgia. They had met

---

Hundreds of men drowned as the holds, crammed full of sick and injured, filled with water. Japanese escorts, after rescuing the Japanese soldiers who had survived the attack, machine-gunned the British and Dutch POWs clinging to life rafts and debris.

while Lawrence was attending Georgia Tech and Sarah was attending Hollins College in Virginia. Their romance continued through Lawrence's graduation from the Naval Academy and Sarah's graduation from Agnes Scott College in Atlanta.

Lawrence and Sarah came from prominent families in Columbus and Atlanta, Georgia. Both were descended from ancestors long associated with the cultural and business life of those cities. Their June wedding became a centerpiece in the society pages of the *Columbus Ledger-Inquirer* and the *Atlanta Constitution*. The *Atlanta Georgian*'s society columnist, Polly Peachtree, called it a "fashionable event," and dubbed Sarah "the charming Atlanta belle" and Lawrence "the handsome naval officer." The papers featured long, detailed descriptions of the bride's and her attendants' gowns, the floral arrangements, the music, even the interior of the church. The *Constitution* gushed that attendees included fashionable members of Atlanta society and prominent out-of-town guests.

A month after their wedding a full-page banner headline topping the *Constitution*'s society page announced: "Pacific Ocean Borders Front Yard of Atlanta Bride." An accompanying article informed society-conscious Atlantans that the newlyweds had arrived in the exotic Pacific outpost of Hawaii, with its palm trees, swaying grass skirts, and uninhibited American sailors, and where Lawrence, now a lieutenant junior grade, started his career in submarines aboard the big V-class *Narwhal* (whose six-inch guns would in the future bombard Matsuwa To during that pioneering foray into the Sea of Japan).

After he completed his tour in the *Narwhal* in December 1940, Edge reported aboard the rusty World War I–vintage submarine *O-4* based at the Philadelphia Navy Yard, where many obsolete submarines were laid up before scrapping. Unlike her sisters destined for the breaker's yard, the *O-4* had been chosen by the Navy for duty as a submarine school training boat. Edge worked hard and put in long hours on the refurbishing and fitting out of the rusty old sub. Then, in July 1941, with World War II looming for America, Edge received orders to report to the Naval Academy to undertake a two-year postgraduate course of instruction in what was then called "radio engineering." This new field married electronics to communications and other technologies that were coming into wider use in the Navy. The equipment required highly skilled technicians to operate it and repair it.

The assignment must have convinced Lawrence that he was doomed to attend classes ashore while the war everyone knew was coming got started without him. He may have doubted that he'd ever get aboard a submarine to join the battle. After all, it was no secret that submarines would play a major role in a war against the Japanese, and every submariner alive, veteran and novice alike, wanted aboard one. A lot happened during the two years he and Sarah spent together in Annapolis. The most important event was the birth of their first child, Sarah. Meanwhile, overseas, Hitler, after conquering eastern Europe, invaded the USSR. The Japanese, after attacking Pearl Harbor, overran the Far East.

Lawrence endured his two years at Annapolis far from the action, studying basic and advanced electronics. His course of instruction included periodic bouts of temporary duty at places like the Sperry Gyroscope Company on Long Island, the National Broadcasting Company in New York City, the Radio Corporation of America in New Jersey, Philco Radio & Television Corporation in Philadelphia, and many others. His duties included visits to the Fleet Sound School in Key West, Florida, and the Massachusetts Institute of Technology in Cambridge. He and a pregnant Sarah moved up and down the East Coast together, often living in temporary quarters while enjoying married life in the Navy even as Lawrence grew impatient for submarine work.

The companies and institutions where Lawrence learned the practical working side of electronics had earlier started gearing up for war by developing and then refining the very technologies Edge was studying. During the period that Lawrence and Sarah traveled up and down the East Coast from duty station to duty station, Manila fell to the Japanese; the U.S. Navy mauled the Japanese Navy at the Battle of Midway; the Allies began to turn the tide against the German U-boat onslaught in the North Atlantic; and Russian troops, after encircling Germany's Sixth Army, began to break the Wehrmacht's stranglehold on Stalingrad.

Lawrence's salvation, as it were, arrived in two phases. On May 1, 1943, he received a promotion to the rank of lieutenant commander. Then in late July, as he was completing his studies, he received orders assigning him to New London for instruction at the Prospective Submarine Commanding Officer (PCO) School. This was a major turning point in his life and career, for the goal of any seagoing naval officer, especially in wartime, is command at sea.

An officer selected for PCO School and possible command of a fleet-type submarine faced a formidable challenge. This despite the fact that when the United States entered World War II the Navy needed all the trained submariners it could get to man the new boats coming out of the building yards at the rate of four or five a month. Only the very best officers received training as potential skippers. To succeed in the pressure-cooker environment of submarine combat, a candidate for command had to be not only highly intelligent and have an analytical mind, but he also had to be resourceful, self-confident, and above all, unflappable under pressure. Consequently, PCO candidates were carefully screened to weed out individuals who lacked the qualities essential for command; a lot of men didn't make the grade. Men like Lawrence Edge, who did, joined a select group destined for command of a submarine in a Navy that had undergone such swift and far-ranging changes in outlook and structure that it scarcely resembled the Navy America had when it entered the war.

With the increased tempo in the Pacific at last turning back the Japanese, Lawrence arrived in New London in August 1943 for PCO instruction, which he completed in late September. He then received orders to report in late November to the Seventh Fleet and to the headquarters of ComSubSoWesPac, 13,000 miles away in Fremantle, Australia.

Lawrence took leave at home with his family in Atlanta before departing for Australia. He and Sarah knew that they'd not see each other again for a long time, perhaps a year or more. Though eager to undertake his new duties, which would culminate in command of a submarine, Lawrence knew that this achievement, important as it was, wouldn't ease the pain of prolonged separation from his wife and daughter and the longing it would cause. It was no different for Sarah and for all the wives, husbands, and friends of the men and women serving in the armed forces; they, too, would suffer the pain of extended separation, and with it the constant, nagging fear that their loved ones might be killed.

As for submariners, the fear of death was something they learned to live with, though it was never far from their minds, given the dangers inherent in submarine operations themselves, and not just from enemy action. Yet by their nature submariners are optimists. They have to be, for after all, they'd volunteered (the submarine service is an all-volunteer force) to be sealed up in a steel hull—some would say an iron coffin—for the duration of a war patrol. Other than the ship's officers only a handful of enlisted men ever got to see the world outside their submarine during a patrol. What kind of men would volunteer to spend a good part of their lives living together in cramped, hot, smelly spaces enduring physical and mental stress while under constant danger from the Japanese, if not their own machinery and the sea itself?

One answer is that the submariners' lifelong bond of brotherhood and camaraderie with shipmates is like few others in the military service. Another answer is that because the duty is so demanding submariners have always had a certain mystique, as though schooled in some black art or arcane specialty, which is absolutely true, given how complicated subs are. Aboard a submarine, more than in most warships, each man is dependent upon his shipmates for the sub's performance and safety, if not his own survival. Caution and vigilance are the watchwords aboard a submarine prowling beneath an unforgiving sea, where one mistake can lead to disaster. Thus, after undergoing rigorous and highly specialized training, each officer and enlisted man strove to earn the designation "qualified in submarines" and to wear the twin-dolphins insignia that marked him as a man apart, a man belonging to an elite service, a man who was special, even fearless.

Fearless indeed. Up until just before the start of World War II, it was dangerous to go to sea in a submarine, much less submerge in it. The old submarines of the 1920s, the O-, S-, and R-class boats, though vastly improved over the subs Charles Lockwood had once served in, were still nothing more than leaky rust buckets. Often their hatches didn't seal properly, and as seawater sluiced into the boat through inch-wide gaps in the hatch's knife edge, sub crews could only hope that sea pressure would eventually seal them, which it usually did, but not always. Then there was the stench of sweat, oily bilges, and stopped-up heads brimming with human waste. Back then sub sailors had to endure one-hundred-degree-plus temperatures, lack of air-conditioning, and choking diesel fumes. Things weren't much different from what they were in Lockwood's day. Stalwarts said that conditions like these built character and fostered camaraderie. No doubt they did, though with the advent of the modern fleet-type submarine in the late 1930s, conditions vastly improved for sub crews. More than shared hardship, the camaraderie fostered during World War II resulted from the shared experience of men fighting to defeat a vicious and pitiless enemy. The camaraderie that such experience engendered in submariners patrolling the Pacific Ocean was outside Sarah's understanding. She knew only that Lawrence was committed to his profession and that he was determined to carry out his duties to the best of his ability. As close as she and Lawrence were, she could never share with him the closed world he was about to enter. When they parted early in November 1943, there was no way for either of them to know whether they would ever see each other again.

Edge departed from the West Coast on a lengthy sea voyage "down under," to Australia. From Brisbane, he flew cross-country in an Australian National Airways DC-3. Days later, after arriving at the sprawling submarine base in Fremantle, he reported for duty to Submarine Squadron Sixteen.

The submarine base at Fremantle, on Australia's western coast, bustled with activity. Submarines departed almost every day on war patrols, while others returned from patrols for refits and crew rotations. A sense of urgency permeated every aspect of submarine operations, from the staff level down to the sailors chipping rust on the boats. Swept up in this urgency, Edge, with little time for rest after his long trip, reported to Admiral Christie's ComSubSoWesPac operations staff to undergo a two-month crash course in submarine operations, logistics, ordnance, personnel, and much more.

Then, in late January 1944, in preparation for taking command of a submarine, Edge received orders to the USS *Bluefish* (SS-222) as executive officer. It was invaluable experience for a PCO, as an exec has responsibility for the day-to-day operations of the ship, including all the paperwork and personnel headaches that go with it, to say nothing of his duties as navigator and, in some boats, fire control and attack coordinator. Edge served in the *Bluefish* during two war patrols under two skippers: Commander George E. Porter and Commander Charles M. Henderson. They could not have had a better exec than Edge.

Sarah always received a cable from Lawrence upon his return from a war patrol ("ALL WELL AND SAFE. LETTERS RECEIVED . . . FONDEST LOVE AND KISSES"), which was then followed by a letter, in this case, one typical of those he wrote during the war.

My most precious, most wonderful, most lovely wife,

Back again [from patrol] to the greatest thrill I know . . . getting once again your wonderful letters. Angel, you can't know . . . the feeling, the soaring to the skies of my spirits upon receiving one of your beautiful letters.

. . . I can think of only one thing which can be yet more joyful to me, and that, I know your heart will tell you, is seeing you and little Boo once again. That, I hate to dwell on for any length of time at all, because it just makes me unhappy about being out here. . . . What I meant is that it makes having to stay out here one minute longer almost unbearable. . . . [H]ow miserably I failed to write you at all this whole last patrol, long as it was.

The worst of it, darling, is that I hardly know how to offer you an excuse. It just seems to me even looking back on it, that I was tired the whole time. We started with a pretty strenuous training period, which tired every one to begin with and continued the training for some time after departing; perhaps I just never quite caught up. We did operate somewhat differently as compared to the previous patrol, due principally to the new skipper [of the Bluefish*] with different ideas from Capt. Porter. We operated submerged much more of the time and no one feels quite as chipper as when on the surface in the fresh air and sun.*

. . . Anyhow, I still consider myself very lucky indeed to have had the training I have had with these two patrols under two different skippers, each of whom I consider good, different as they are. Yes, their very difference has been an advantage to me, being worth more, I am sure, than two patrols with the same one—not that I learned all there is to learn from either one, of course.

The skipper of the next boat I'm on won't be so easy for me to analyze, however, as in the case of the Bluefish. Yes, I've made my last patrol on the good old Bluefish—and now I'm ready, perhaps, for the first time, for my next job. The new job I'm promised, is to begin soon, which really delights me. I would dread a long wait, as so many of the boys have had. I don't yet know for sure which boat it will be, but the sooner the better.

When I first came out here, as you know, I wasn't really sure whether I was ready for this next job; at the end of my [first job] . . . I felt better, but still not too sure; now I do feel sure that I'm ready for it. No, I don't feel overconfident, but just that the next step in my education for the job will be the job itself. Anyhow, I pray to be truly worthy of it, because if I can be truly successful at it of all jobs, I'll be able to feel that at last I'm really doing something to hasten the end of this war, and my return to you and Boo—which is all I really live for now. Each ship we sink seems to me to cut the long wait for that great day by another hour, or perhaps day, or week, or month, and I am glad (though I hate to think of what has happened to some of the poor mortals, Jap though they be, who happened to be on some of those ships).

I'm not sure how I got started on my "philosophy of war," or

lack of it there, because what I was really discussing was your
indescribably wonderful, faithful, and beautiful letter.

 . . . Anyhow, in my next job I hope to . . . [write] as nearly
as possible each and every day, at least a few lines.

  . . . [With] all my dearest love, sealed with all my kisses,
for you and little . . . Boo.

Lawrence[2]

As recounted earlier, Edge relieved Hogan on June 13. In only eleven months, and with New London, Annapolis, and Atlanta on the other side of the world, Edge found himself standing on the *Bonefish's* quarterdeck taking his crew's salute and being addressed as "Captain."

On June 27, just hours before the *Bonefish* departed on her fifth war patrol, Edge started a letter to Sarah, telling her how lonely he felt, that he was suffering from a bad case of the blues, adding:

Dearest, most precious love . . . maybe the time will really come
some day when this is all over, and I'll be holding you and
[Boo] in my arms again! Oh, Shug, I shiver all over to think of
it, it's so wonderful.[3]

The next day he completed the letter, in which he hinted that he now had command of his own sub.

Dearest Angel, I do hope this patrol will not find me too busy or
too tired to talk to you often in letters. Last patrol was no fun,
especially on that account. Even though I don't like the mental
discipline . . . required to put my thoughts on paper, I do like
talking to you, sweetheart, better than anything I can do out
here, except reading and receiving letters from you.

 So far, I like my new job fine, and if we make out as well as
I'd like to this time, I'll definitely call it superior to any I've ever
had before on a ship! So keep your fingers crossed and don't
forget to say a prayer for us! . . . And don't worry about us
either, sugar mine; this may be another long [war patrol], and

perhaps even longer [than others]. As to length it is utterly
impossible to tell in advance, as you know. My last one [in the
Bluefish] was actually just normal full time. Only
exceptionally good hunting or damage, or other casualties
[would] cause them to be shorter. The long ones, then, are
really the rule and the short ones the exceptions. If ours this
time is another long one, just remember that I'm thinking of
you and Boo just as much as you of me, to say the least, and am
anxious to get back to receive your letters. . . .

After topping off provisions and fuel at Exmouth Gulf, a forward base on
the coast of Australia, the *Bonefish* headed for her patrol area in the Cele-
bes and Sulu seas north of the Malay Barrier. U.S. submarines had found
good hunting in these waters. Despite questionable torpedo performance,
SoWesPac subs had sunk scores of Japanese *marus* along with their cargoes destined for the home islands. They had also sunk scores of other
marus doubling as troop transports. Thousands of Japanese soldiers
along with their weapons, ammunition, and supplies had thus failed to
arrive at the many Imperial Army garrisons scattered throughout the
conquered territories in East Asia. For the Japanese the loss of these
needed troop replacements and food supplies was like dying a slow death
from a thousand bleeding wounds.

Early in the patrol the *Bonefish* encountered fleets of native fishing
vessels and other small craft. For a while Edge shadowed a pair of sea
trucks—small wooden-hulled cargo haulers—thinking to sink them
with the sub's four-inch deck gun. But before Edge could act the two gave
him the slip into the shallow coastal reaches of Lombok Island. No matter, Edge sought bigger game.

He found it on July 8, when the *Bonefish* picked up a convoy of four
escorted ships making a sortie through an interisland channel off Cape
Mangkalihat. Edge made a high-speed submerged end around to reach a
position ahead of the convoy, a maneuver that required nearly five hours
to complete. As the *Bonefish*'s tracking party followed the convoy's movements, they input fire-control data—the convoy's range, bearing, speed,

and other essentials—into the sub's Torpedo Data Computer (TDC). A mechanical analog computer, the TDC solved the complex mathematical equations associated with firing torpedoes from a moving submarine at a moving target and hitting it. The TDC didn't guarantee success; it just made the job easier.

After sidestepping a group of slow-moving single-masted luggers, Edge swung the *Bonefish* around to fire torpedoes from the stern tubes. If the attack proved successful and all hell broke loose in the convoy, it would cover the sub's withdrawal. If a follow-up attack was necessary to finish the job, he'd swing around to fire torpedoes from the bow tubes before the convoy broke up and fled.

The TDC's angle solver and position keeper hummed and whirred as the *Bonefish* closed in. When the machine's "correct solution" light flashed green, the executive officer, in charge of fire control, bawled, "Shoot anytime, Captain!"

Edge motioned with both thumbs. "Up scope!" A quick look revealed that the ship he'd chosen to torpedo would cross the *Bonefish*'s stern in about a minute. The TDC said that he had to shoot now or else lose it.

"Fire . . . !"

The *Bonefish* lurched once, twice, three times as her torpedoes lunged for the target.

Edge saw three streaking wakes emerge from the *Bonefish*'s stern even as the soundman, headphones clamped to his ears, reported hearing the tin fish running hot, straight, and normal. Edge and his crew waited expectantly, if not patiently, for the torpedoes to reach their target.

"Captain, torpedo noises mingling with target screw noises."

Edge looked at the passing parade of darkened ships and saw . . . nothing. The *marus* and escorts continued merrily on their way, apparently unaware that they had escaped being torpedoed.

Edge didn't hesitate a beat. After a quick TDC setup he swung the *Bonefish* around, fired five torpedoes from the bow tubes at a second target, then, rudder hard over, veered away from the convoy. Five bubbling torpedo wakes spread like fingers, headed for the target. An escort spotted them, heeled around, and, stack belching black smoke, bow cutting water like a knife blade, barreled straight in after the *Bonefish*. It was time to dunk the scope and go deep! *What about those fish?* Still nothing. The escort's screws beat a steady inbound rhythm until her skipper veered away, wary of taking a sixth torpedo in his snout.

Running deep, the *Bonefish* evaded detection. Meanwhile, before Edge could regroup for another attack, the convoy hightailed it out of sight. All he could do was shake his head: eight torpedoes fired with nothing to show for it.* Edge reported that he and his men were "feeling mean enough . . . to bite any Japs on sight."

Ten days later, patrolling east of Palawan at midnight, *radar contact!* The officer of the deck (OOD) reported the target's vitals: range, bearing, speed. "Station the tracking party," Edge ordered.

He started the pursuit on a course parallel to the target, then slowly closed in to identify it. A night dark as ink, with heavy overcast and fast-moving rain squalls, made a visual identification impossible, though the size and strength of the target's radar blip indicated it was worth a torpedo.

While the tracking party kept the TDC updated with radar ranges and bearings, the chase unfolded across miles of ocean. Time dragged. Two hours after the initial contact, the *Bonefish* arrived at a favorable firing position, though Edge still hadn't seen the target. No matter; he had the setup pictured in his mind—the target's course, angle on the bow, distance to the track, all of it.

Battle stations torpedo!

As the *Bonefish* closed in, Edge ordered: "Make ready all tubes! Open outer doors. Stand by forward! Stand by One."

The sailor facing the torpedo firing plungers numbered One through Six stood ready at Edge's order to slap them home.

---

*Unknown to Edge, his second torpedo shot sank the *Ryuei Maru*. On March 18, 1946, the office of the Chief of Naval Operations released a document entitled, "Reassessment of Damage by Submarines." In it, the CNO stated that, regarding the *Bonefish*'s attack on a convoy in the northern approaches of the Makassar Strait on July 8, 1944, "Japanese intelligence states that the *Ryuei Maru* was sunk at this time. Tonnage of that vessel is not definitely known but is estimated at 2,300 tons."

Once again the exec at the TDC with its clicking machinery reported, "Shoot anytime, Captain."

"Fire One!"

In quick succession three torpedoes howled out of their tubes. Two veered off on erratic runs. *Damn the tin fish!* There was no hope for them; they were long gone who knew where? A stopwatch thumbed by the exec timed the sixty-second run of the third straight-running fish. Sixty seconds dragged like sixty hours.

Edge, on the bridge peering into a seemingly impenetrable curtain of ink, had almost given up hope for the third torpedo. Then a bright flash followed by a sharp boom signaled success. After ten straight misses Edge had the satisfaction of seeing and hearing a *Bonefish* torpedo blast a hole in the hull of an enemy ship. He watched it sink under an oil slick and a vast mat of floating wreckage.

After the attack the *Bonefish* maneuvered carefully through a sea covered with lumber, straw mattresses, capsized and wrecked lifeboats. In the middle of this mess she nosed alongside a group of twenty-five oil-soaked survivors clinging to debris. Somehow Edge coaxed aboard the sub one of the men too exhausted to swim away. Japanese rarely submitted to being taken prisoner at sea, preferring to drown themselves than undergo the dishonor of capture by the enemy. In passable English the man claimed he was the ship's boatswain and that she was bound from Negros to Manila with a crew of thirty-two plus 124 naval ratings and six officers, most of whom had drowned. Edge cleared the area, keeping the prisoner for interrogation by Navy intelligence back at Fremantle.

The night of July 30, Edge attacked another convoy. This time the escorts, *Chidori*-type torpedo boats and a destroyer, detected the *Bonefish* and held her down long enough to allow the convoy to escape. If they thought Edge would slink away, they were wrong.

Edge chased the convoy and caught up with it. Working in from a parallel track, he allowed the leading starboard escort, the destroyer, to go on by. Then he cut in behind it to fire torpedoes at the main convoy body. On the bridge Edge saw four evenly spaced torpedoes explode against the hull of an oil tanker. "Four beautiful hits seen, heard, felt, and timed [Edge reported], with two equally forceful internal explosions likewise recorded. . . . Near perfect torpedo performance if ever we had seen it."

Edge hauled out as the escorts began dropping depth charges on what they thought was the submerged *Bonefish*. "Felt sorry for any survivors in the water. Four more depth charges; they sound much better [when we're on the surface] than down below."[4]

In between chasing and torpedoing ships, the *Bonefish*'s gun crews shot up and sank five miscellaneous luggers and sailboats. Edge had orders to be on the lookout for any vessels doubling as submarine spotters. These were usually large twin-masted sailboats with radio antennas strung between their masts. Though these and other small craft often looked more like innocent fishing vessels than spotters, it was hard to tell from a distance. Overhauling and boarding them was risky work. The armed *Bonefish* boarding parties had no way to know whether their crews were friendly Filipinos, Malayans, or Chinese, or whether they were Japanese, armed and ready to put up a fight.

Typically the *Bonefish*, her deck guns manned, approached a suspect boat, circling it as the boat's crew doused sails in preparation for being boarded. Sometimes it took bursts from the sub's .50-caliber machine guns to convince them that Edge meant business. In a battle-surface encounter with a large, motorized two-masted schooner that refused to heave to for boarding, the *Bonefish*'s gun crews laid down a barrage that splintered the boat's hull and deckhouses and blew up a load of fuel drums. Some boats fought back. A pair of small wooden cargo vessels returned fire with a machine gun, bullets zinging off the *Bonefish*'s hull. Another vessel, though holed by four-inch rounds, tried to ram. The *Bonefish*'s gunner's mates made short work of it, pumping in round after round until, with its topsides ablaze, the vessel drifted away to sink.

Edge soon discovered that among all the small craft plying the waters south of the Philippines, few were spotters. Most were crewed by friendly Filipinos fleeing the Japanese. Those who spoke a little English were eager to provide Edge with information about Japanese ship movements in the area. In return, Edge, with little in the way of provisions to spare, gave them fresh water, packs of cigarettes, a little food, and in one case nine hand grenades from the *Bonefish*'s armory.

Edge reported that later in the patrol,

> [O]ne of the vessels [we] stopped looked harmless at close quarters, with only brown-skinned crew showing. The complement consisted of eight men. Two women, and one baby (prominently displayed, age about eight months). They hailed as best as we could make out from Macassar, no English spoken. She carried no cargo other than own food and water, and a few boxes of Javanese cigarettes made at Soerabia. A check disclosed considerable money in the form of Japanese-issue guilders, and a ledger indicating that the boat had called at Macassar, Soerabia, Batavia, and Tarakan. Upon our approach the crew had not hidden, but lowered sail without any show of force on our part and willingly helped bring the boat alongside by tending lines. They appeared friendly in every way. Wished we had something to leave them as a parting gift, but [there was] nothing we could spare, whereas we were offered both cigarettes and money.[5]

The *Bonefish*'s crew and the Filipinos waved good-bye and departed friends.

Midday August 2, as the submerged *Bonefish* eased south along the Zamboanga Peninsula of Mindanao, the masts of three ships escorted by circling airplanes lumbered into view. They turned into a five-ship convoy consisting of one large empty tanker, a medium-to-large freighter, and three smaller ships providing escort. *Battle stations submerged!*

Edge tracked the targets, careful to minimize periscope exposure to avoid detection by the circling planes. He bored in on the tanker's starboard flank, confident that her escorts and the circling planes hadn't spotted the stalking *Bonefish*. He swung right, fired three stern tubes from very long range, 1,700 yards, then swung left, ready for a bow shot if needed. Stopwatches timed the runs. Sonar reported hearing only two fish, not three. Edge just shook his head. But an explosion 1,600 yards into the run signaled success—or was it premature or a bomb dropped from one of those circling planes?

"Up scope!" Edge reported a solo hit on the tanker. It had damaged her, but not enough to sink her. Once again erratic torpedo performance had marred a successful attack.

Edge spun the scope for a quick look around and saw the panicked escorts rushing to and fro, an indication that a hunt for the *Bonefish* was under way. But it proved a halfhearted one that quickly ended as the scattered convoy cleared the area.

Determined to finish off the tanker, Edge started an end around to get out in front of the convoy to attack as it plowed south. The end around chewed up almost thirteen hours and it wasn't until two the next morning that radar picked up the convoy disposed in a ragged column paralleling the Zamboanga coast near the Basilan Strait.

The tanker Edge had damaged earlier, her unmistakable silhouette masked somewhat by the dark land background, had two shepherding escorts stationed port and starboard. Edge waited until the moon set behind thick clouds; then, with the *Bonefish* flooded down, decks awash to minimize her silhouette, he came in fast on four roaring engines, wary that the *Bonefish*'s bow wave glowing white in the phosphorescent sea might be spotted by an alert lookout on the outriding escort.

Edge ran the *Bonefish* parallel to the target's track while he waited for the escort to go by. When it did, he swung the sub's nose toward the tanker.

Edge directed the attack from the bridge, where he could see the ships, dark, blocky shapes darker than the Zamboanga peninsula behind them. He bawled orders into the bridge speaker to the attack team below in the conning tower. "Left full rudder!" The *Bonefish* heeled onto a new course. "Steady as she goes!" Arrow straight, bow pointed directly at the tanker's side looming up ahead, the *Bonefish* bored in.

"Fire One . . . ! Fire Two . . . ! Three . . . ! Four . . . ! Left full rudder! All ahead flank!"

Torpedoes fired, the partially awash *Bonefish* rose from the sea as her howling low-pressure blowers forced water out of flooded ballast tanks. Retrimmed for high-speed cruising on the surface, she wheeled around and pulled away. In her wake four white bubble trails converged on the tanker.

Looking aft, Edge counted down the seconds. The fish should be there by now. . . . A miss! Another miss! And another! . . . *What about number four . . . ?* He saw a small flash erupt at the tanker's waterline, then a white geyser taller than her bow. A hit! A signal lamp on one of the ships began to flash an indecipherable message to the rest of the convoy. If it was a warning, it came too late.

Edge had little time to savor his work. An aircraft contact on SD radar forced the *Bonefish* down. Passing a hundred feet, the sonar watch reported a distant explosion—one of the torpedoes that had missed the tanker had probably exploded when it hit the beach or sea bottom at the end of its run. Edge hoped to hear breaking-up noises from a sinking tanker. Instead, as the explosion faded away, sonar reported only distant screw noises and long-range depth charging.

Edge was deeply disappointed by his failure to sink the tanker, not just for himself but also for his crew. He was painfully aware that all of those tension-filled hours of tracking, plotting, and sweating had resulted in damage to only one ship. He could have blamed his failure on erratic torpedo performance that continued to curse the *Bonefish*. But Edge being Edge, he was as tough on himself as he was on the enemy. He vowed to make amends the next time he had a Japanese ship in his sights.

The next day the *Bonefish* ran through a large area of floating wreckage—oil drums, lifeboats, splintered lumber, life rings—possibly from one of the ships she'd torpedoed earlier or from another victim of a U.S. submarine that had been in the area before the *Bonefish*'s arrival. The coast of Mindanao had become a graveyard for Japanese *marus*.

Patrol at an end and headed home via the Lombok Strait east of Bali, the *Bonefish* exchanged recognition signals and information with the northbound submarine USS *Flier* (SS-250). The *Bonefish* was likely the last submarine to make contact with the doomed ship. On August 13 the *Flier* struck a Japanese mine while transiting the Balabac Strait south of Palawan and sank in less than a minute. With help from Philippine guerrillas eight *Flier* survivors, including her skipper, were rescued by the USS *Redfin* (SS-272).

Admiral Christie noted in his endorsements to the patrol report Edge submitted after his return to Fremantle that the new skipper had conducted the patrol in an aggressive and thorough manner. He concurred with Edge's assessment of sinkings and damage: one freighter and one oiler sunk, another oiler damaged. He noted too that Edge's report was well written and filled with operational details on Japanese antisubmarine tactics and convoy routines that would be useful to other skippers assigned to patrol the same area.

Edge's narrative provided Admiral Christie with an exceptionally clear and accurate picture of the action that had unfolded day by day, hour by hour, minute by minute aboard the *Bonefish*. Christie surely found the report, with Edge's colorful personal observations inserted into the narrative, as lively and exciting as a fast-paced short story. For his outstanding performance, Edge received the Bronze Star, the *Bonefish* her fourth unit commendation.

In an undated letter written during the long, arduous days of patrolling in enemy waters, Lawrence told Sarah in words carefully chosen to avoid censorship that he'd just concluded his first patrol as skipper of the *Bonefish*. He also dropped a hint that before too long the *Bonefish* might return to the States for an overhaul. He also explained his reasons for fighting in a war that seemed it would never end.

> *Dearest, most lovely of all sweethearts, I love you, I love you, I love you. . . .*
>
> *Again a patrol is nearly over . . . there are things I want to tell you—so many that the censor will not let me tell—and most of*

the others just adding up to the fact that I love you two and miss you so terribly much that I can hardly bear dwelling on it. . . . What we are doing each day is, of course, our part in the war, be it big or small, and as such I can't tell you. The nearest thing to a diary allowed on board is the . . . patrol report.

Just recently, I've read a little article in last January's Readers Digest, a letter of an Army doctor to a friend telling . . . his thoughts, feelings, and experiences during 60 days of the heaviest fighting in New Guinea. Luckily, I have not had to go through all the terrible experiences he had in that bitter jungle warfare, but much of what he says I know to be true— and none more so than his last paragraph:

"But most of all I know that the best thing on earth is the love of a man's wife, and the sustaining strength of a man's family at home."

In the last analysis, I'm sure that that is what most of us are really fighting for in our hearts—for our country, the place where our wives and families are, the place which we want to keep safe and happy for you, so that we can eventually return to you there and live the kind of life with you that both you and we believe is the best the world has yet to offer. . . .

As for me . . . I've won no medals this time, certainly [his Bronze Star had not yet been awarded]. *But I'm not complaining very bitterly. Our luck in many ways was very good indeed . . . insofar as damage done to the enemy is concerned. Yes, it has . . . earned another star for my combat pin. In fact I guess it was somewhat more successful than the last patrol I was on. And I'm not nearly as tired out or whipped down this time as last, either. In practically all respects, too, my job this time has continued to be the best one I've yet had in the Navy. But that doesn't mean I won't really be glad when this war is over and I can be with you two once more. . . .*

Lawrence concluded with news he'd heard that the war in Europe might be over by early 1945. And he took as a sure sign that the Navy's drastic reduction of its submarine building program meant victory in the Pacific wasn't too far off either. It also meant that Lawrence would probably not receive a new-construction submarine of his own to command, which every combat skipper hoped to achieve. Yet as he always did, Lawrence delighted in the bounty of his life instead of dwelling on his misfortunes.

> Anyhow [the Bonefish] has to go back to the yard [on the West Coast] eventually. That won't be [for] as long or as good [as taking command of a new submarine], but I guess I have no just cause for complaint . . . if that should be the worst thing to befall me.

CHAPTER FIVE
The "Magic" Behind the Mission

I t had been a bitter and bloody fight in the Pacific. American forces had suffered terrible casualties in their island-hopping campaign against the Japanese. Starting in March 1942, the Japanese had completed their capture of Java. In May the last Philippine bastion, Corregidor, fell, and with it the remnants of organized U.S. and Filipino resistance. The Japanese army seemed invincible. Yet their defeat at the Battle of Midway in June 1942 marked a turning point in the war. For Japan, the loss of four aircraft carriers (the U.S. lost one carrier), with their planes and pilots, marked the beginning of the end. Yet even as the Japanese fell back across the Pacific in the face of ever more powerful American advances, Japanese soldiers chose to fight to the death rather than surrender. And though America and her allies had taken the initiative, it was clear that the Pacific war was not going to end suddenly, and that it was going to consume more and more lives on both sides. Worse yet, an invasion of the home islands, a dreaded possibility, would likely require five million soldiers, sailors, marines, and airmen, and incur perhaps a half million casualties.* Even though facing defeat, the Japanese still had the capacity to inflict heavy losses on an invading army. At risk, too, were the thousands of Allied prisoners of war and civilian detainees. Therefore, any military operation that might speed Japan's collapse and lessen the need for an invasion would receive serious consideration and likely gain approval from Admirals Nimitz and King. They had already approved a naval

---

*At war's end deaths among American and British Commonwealth troops in the Pacific theater numbered approximately 108,000.

blockade of the home islands and air strikes on industrial targets across Japan. But no one knew if they would be effective, or, if they were, how long it would take for their effect to end the war.

Looking back after the war Lockwood realized that the operation he'd envisioned for a submarine raid targeted against Japan had had its genesis during the visit he made to the laboratories of UCDWR at San Diego, California, in April 1943, where he'd had his first glimpse of FM sonar. Given his enormous responsibilities and all the issues relating to submarine combat operations vying for his attention, the idea that FM sonar might provide the means to penetrate the minefields guarding the Sea of Japan had taken a while to germinate. When it did it reignited Lockwood's determination to once again make that sea a theater of operation for his subs.

Sometime after the *Wahoo*'s loss in October 1943, an intelligence report arrived at ComSubPac challenging Lockwood's belief that the *Wahoo* had been sunk by a mine in the Sea of Japan. Lockwood had clung to that idea absent an official announcement by the Japanese that their antisubmarine forces had sunk the *Wahoo*. To Lockwood the lack of an announcement by the Japanese indicated that, unknown to them, the *Wahoo* had struck one of their mines. Support for this idea came from submarines operating around Japan, whose skippers reported an increase in the number of floating mines that had broken loose of their moorings and gone adrift, sometimes into traffic lanes used by the Japanese themselves. The dangers they posed to submarines as well as surface ships were all too real, as evidenced by the sinking of the *Flier*.

The intelligence report Lockwood received indicated that Japanese antisubmarine forces had attacked and sunk a submarine in La Pérouse Strait on October 11, 1943.* Why the Japanese hadn't made an official announcement to that effect was a real mystery, given their penchant for exaggerating reports of attacks on U.S. subs. There was no denying that

---

*This was not the report of the attack on the *Wahoo* that U.S. intelligence teams unearthed in Tokyo after the war, as in chapter three.

the submarine in question was the *Wahoo*. Lockwood mulled over what he knew for sure. The *Wahoo* had gone in there twice. Other subs had too and had gotten out. Only the *Wahoo* had been sunk, and apparently not by a mine. If a mine hadn't sunk her, did that mean the Sea of Japan wasn't impregnable after all? Did it mean that the minefields guarding the sea weren't as formidable as Lockwood and Voge had thought? Could they be penetrated—somehow—without incurring the loss of men and ships?

Lockwood realized that UCDWR had unintentionally created the tool that would make it possible to attempt a mission that he believed would keep his submarines in the fight. The war was evolving into a naval air campaign just as the submarine war of attrition against Japanese shipping was slackening for lack of targets. His force was on the brink of going out of business. Lockwood's mission would sink the Japanese empire once and for all, and in the bargain avenge Morton and the *Wahoo*.

A big-picture man, Lockwood envisioned a mission that had three main objectives: One, penetrate the minefields guarding the Sea of Japan to prove that submarines could do it. Two, show the Japanese that they were virtually isolated and defenseless against submarine incursions. Three, cut off the imports of rice, coal, and iron ore from East Asia that Japan needed for survival. Lockwood's submarines would accomplish these objectives by sinking every last ship still afloat in the Sea of Japan. Another important goal coming into focus as a result of America's sometimes difficult collaboration with the Russians, though not articulated by Lockwood, was to demonstrate to the leaders of the Soviet Union that the United States Navy's powerful submarine force would have a role to play in the implementation of America's strategic objectives in the postwar world.

In early 1944, after several follow-up visits to UCDWR's labs to see what progress had been made on FM sonar, Lockwood came away convinced that the new device had the potential to become the secret weapon the sub force needed to wipe out the remains of Japan's merchant marine. He was more determined than ever to get the sonar units into full-scale production and to get one installed in a submarine for trials as soon as possible.

Lockwood outlined for Admiral Nimitz what he'd learned about FM sonar. Impressed by what he heard, Nimitz approved Lockwood's request to have the first available sonar unit installed aboard a submarine. His approval also allowed Lockwood to set in motion an almost continuous rotation of civilian scientists and instructors between UCDWR in California and the sub base at Pearl Harbor and, later, at Guam. At both California and Pearl Harbor, the scientists would supervise the installation and repair of FM sonar units in subs, and the training of the submarine personnel who would operate them. This cooperative effort would give the scientists an opportunity to experience the hot, humid, and rugged environment of a submarine, where the sonar equipment would have to operate without malfunctioning. It would also provide an opportunity for the submariners undergoing training to become acquainted with the difficulties inherent in the development and manufacture of extraordinarily complex and temperamental equipment, as well as its operation and maintenance. From the beginning, and despite all the problems imposed by rush schedules, stress, short tempers, and nagging production bottlenecks, the collaboration between scientists and submariners turned into a successful partnership that continued beyond the end of the war. Lockwood couldn't have been more pleased, then, when the first working FM sonar unit showed up in Pearl Harbor in June 1944 installed aboard the USS *Spadefish* (SS-411).

The *Spadefish* was a new-construction boat out of the government building yards at Mare Island, California. Her skipper was a veteran ship sinker, the mustachioed Commander Gordon W. Underwood. After her fitting out, shakedown, and crew training at Mare Island, shipfitters installed a handcrafted FM sonar chassis, sonar head, and associated accessories in her forward torpedo room. After installation the *Spadefish* moved south to San Diego, where UCDWR technicians installed and tested the FM sonar unit's electronics package. Lockwood, anxious to know how the work was progressing, received regular updates on the installation and also on the effort by technicians to rid the *Spadefish*'s unit of the bugs that kept it from performing as it should. Though these problems were eventually solved, they foreshadowed what was to come.

The physical installation itself consisted of several components. They included two soundheads, one of them a transmitting projector, the other a receiving hydrophone. Both were mounted on a rotating, retractable shaft enclosed in a thirty-one-inch-long, twelve-inch-diameter rubber sleeve filled with castor oil.[1] In the *Spadefish*, the shaft from the forward torpedo room ran through the ship's pressure hull on the port side up to the main deck. She was the only sub to have a deck-mounted unit. Later, experience dictated that FM sonar soundheads mounted below a submarine's keel provided better signal propagation out ahead of the sub, which enhanced the sub's ability to locate mines. It also allowed for the use of FM sonar when the sub was running on the surface.

Another component consisted of a four-foot-tall equipment stack inside a steel box mounted to the forward torpedo room deck against the ship's curved hull. The stack contained an FM oscillator, a power amplifier, a receiver, and an analyzer. Another component, a hoist-training mechanism, turned, raised, and lowered the soundhead. A plan position indicator—or PPI scope—mounted in the sub's control room had a circular cathode ray tube similar to a radarscope. Like a radarscope, the PPI scope had a long persistence screen that inhibited image ghosting, a sweep-around circuit, and the necessary power controls to turn it on and off. Distances and relative bearings necessary to navigation were scribed on the face of the PPI scope. A loudspeaker mounted in the conning tower above the FM sonar operator's position sounded a bell-like tone that warned of impending contact with mines. There were also various junction boxes and the necessary connective cabling between the conning tower and forward torpedo room.

Aboard a sub feeling its way through a minefield, FM sonar displayed the returning sonar echo from a mine contact on the PPI scope as a bright green spot of light shaped like a pear.[2] The sharper the pear, the closer the mine. The pearlike display was augmented by the aforementioned bell tone, the volume and clarity of which were directly proportional to the distance from the sub to an actual mine. The bell tone, dubbed "hell's bells" by submariners, and the display of green pears gave the sonar operator a bird's-eye view of the position and range of each contacted mine relative to the submarine. Unlike standard sonar gear, with its individual and discontinuous pulses of sound that often required up to eight minutes for a full 360-degree scan, FM sonar, with its continuous modulated signal, could conduct a 360-degree scan in only eight seconds. In addition, FM sonar's ability to rapidly sweep an area for targets made it difficult for the Japanese to detect its pulse via their conventional single-channel listening gear.

In tests FM sonar sweeps from as far away as eight hundred yards gave good returns from dummy mines in the form of bright green pears and clear, pure bell notes, while poor returns from, say, kelp or schools of fish usually gave off an indistinct green pear and a mushy bell tone, a result of their indefinite shapes. The combined visual display and audio warning allowed a sub to find gaps between rows of mines swaying at the ends of their anchor cables. Or so the theory went. The trick was to thread this forest of mine cables while submerged, no easy feat. After all, mines that are surrounded by cubic miles of seawater are relatively small objects that give off correspondingly weak echoes that can be masked by sound reverberations caused by shallow water, an uneven seabed, and, up above, rough seas. Tests conducted on dummy mines and on triplanes—underwater devices equipped with three perforated sound-reflecting "wings"—revealed just how temperamental the gear could be and how much time and patience it took for the UCDWR scientists to fix problems both mechanical and electronic.

While the *Spadefish* was outfitted with FM sonar, her sonar men underwent special training to learn how to operate the gear and how to interpret those green blobs and bells: Were they mines, kelp, or fish? If they got it wrong, the *Spadefish* wouldn't stand much of a chance if she got tangled up in a row of Japanese mines.

As for her crew of motor macs, electrician's mates, and the like, they weren't impressed with FM sonar and its purported ability to locate mines. So far it hadn't worked as advertised, when it worked at all. And anyway, hunting mines was a job for minesweepers, not submarines. Things went so badly with it that the equipment finally had to be removed for repairs at UCWDR's lab, after which it was delivered back to Mare Island, where it caught up with the *Spadefish* completing her fitting out for war patrols. As scuttlebutt spread through the sub force that more submarines were going to be equipped with FM sonar, sub sailors took a dim view of what might be in store for them. Back in Pearl, Lockwood sifted through the daily reports submitted by Dr. Harnwell, Dr. Henderson, and by SubPac's liaison officer in San Diego, who had his ear to the submarine grapevine. Lockwood discovered just how controversial this new gadget was and how little enthusiasm for it there was among submarine sailors. Lockwood knew that he had to convince both officers and enlisted men to get on board with FM sonar or there'd be no mission to the Sea of Japan. The only way to do that was to perfect the damned thing and then prove that it worked. If he couldn't do it, he'd never put together a task force of submarines to tackle the job he had in mind.

The *Spadefish* arrived at the submarine base at Pearl Harbor in late June. No sooner had she tied up than Lockwood and Voge crossed the brow for a conference with Underwood. The two senior officers peppered the skipper with questions about his experience with FM sonar in tests at San Diego.

Underwood's report wasn't particularly encouraging. Tests of the *Spadefish*'s equipment conducted on dummy mines off the coast of California had been disappointing. FM sonar had failed to register mine contacts visually on the PPI scope or audibly by hell's bells. Sometimes the unit couldn't detect targets at all or had a hard time differentiating between solid objects and schools of fish. It wasn't reliable, said Underwood. Vacuum tubes burned out and wiring overheated. Repairs often took hours. If Lockwood expected submariners to trust this gadget when it came time to locate real mines, he had another thing coming. This wasn't what Lockwood and Voge wanted to hear, but they took the news in stride. Lockwood the optimist believed that he, his men, and the scientists at UCDWR were pioneers in the business of submarine mine hunting and, like any new venture that relied on unproven technologies, it would take time to flush the bugs and gremlins out of the sonar units coming from UCDWR's labs.

In the event, UCDWR technicians who had gone on ahead to Pearl to meet the *Spadefish* gave her FM sonar a thorough going-over. They also supervised some needed voyage repairs—heavy seas had bent the deck-mounted transducer's shaft—after which they pronounced the system ready for duty.

Lockwood put to sea in the *Spadefish* on July 13 to experience first-hand a full-scale test of FM sonar, not on a live minefield but on a dummy minefield planted by the Navy's mine force in the deep waters off Barbers Point, Oahu.

Employing recent intelligence about Japanese mines and minefields collected by JICPOA (ICPOA had been renamed Joint Intelligence Center, Pacific Ocean Areas), the mine force had sown the dummies to replicate an enemy field. Lockwood believed that submarines attempting a penetration of the Sea of Japan would encounter such fields and that familiarity with their layout would provide the submariners the experience and confidence to make a clean run through them.

The Japanese employed a moored spherical, horned contact mine known as a Type 93. Weighing 1,500 pounds, it was filled with 220 pounds of TNT plus a pusher, usually powered aluminum, to increase the mine's explosive force. Later models had up to nine sulfuric acid-filled horns protruding from the mine's cast-iron surface. When a ship hit a horn, it would break open, releasing acid, which energized a battery and set off the detonator. Japanese mines were usually sown in two or three rows across shipping lanes or, in the case of the Sea of Japan, across straits of entry, the rows spaced four hundred to a thousand yards apart. A gap of seventy-five to a hundred yards separated each mine from its neighbors to prevent an explosion from setting off the other mines in the string by chain reaction.

The mines making up a field were generally planted no deeper than a hundred feet, often in three tiers ranging from ten feet to forty feet to seventy feet. A minefield's effectiveness is reduced somewhat by "mine dip," which occurs when deep-water currents push cable-moored mines in the direction of the current flow several feet lower than their planted depth. This so-called "dip gap" gives passing ships a greater margin of safety than they would otherwise have had. Such was the case when Ray Bass ran the *Plunger* through La Pérouse Strait during that early foray into the Sea of Japan. As Bass discovered, mine dip may have accounted for the fact that he and his crew were still alive and their sub in one piece. Even so, Bass's blind run through La Pérouse was thought to be far too dangerous for any submarine to attempt it again. Experience would prove otherwise.

As for those floaters often sighted by patrolling submarines, international agreements stipulated that mines were to be armed only when the weight and tug on the mooring spindle armed the detonator. Otherwise the mine was supposed to self-sanitize, that is, render itself harmless. No one trusted the Japanese to abide by such agreements, so submarines gave floating mines a wide berth. Moored or not, the assumption was that the Japanese had planted thousands of mines in waters bordering Japan to protect the home islands from invasion from the sea. As long as U.S. submarines operated in those mined waters, Lockwood had to hope that FM sonar, aside from facilitating a raid on the Sea of Japan, would reduce the risks that mines posed to submarines engaged in regular war patrols.

Lockwood's ride on the *Spadefish* was an opportunity to gauge FM sonar's sensitivity and accuracy for himself. UCDWR technicians, some stooped with fatigue from the late nights spent tearing apart, repairing, testing, and reassembling the *Spadefish*'s unit, declared it was fully functional again. Lockwood, and especially Underwood, expected to see a vast improvement over its performance in California.

Approaching the dummy minefield off Oahu, Underwood gave the order to submerge: "Clear the bridge! Dive! Dive!"

Two honks of the Klaxon diving alarm sent the men into action. Lockwood and the four lookouts, followed by Underwood, scrambled down the ladder from the bridge to the conning tower below, the eight-foot-by-twelve-foot horizontal steel cylinder set above the ship's control room. Every inch of space inside was taken up with periscopes, FM sonar gear, radar scope, sonar stand, and the TDC. Underwood, his exec, the quartermaster of the watch, the helmsman, sonarman, telephone talkers, and now ComSubPac shouldered around one another trying to make room.

The OOD, the last man down, slammed the hatch shut and dogged it. "Hatch secured, sir."

The hammer of diesel engines ceased as propulsion shifted to the batteries. A litany of orders and repeat-backs resounded between the conning tower and control room. "All ahead full! Rudder amidships! Five-degree down bubble!" In the control room the chief of the watch promptly closed the outboard and inboard engine exhaust valves and main induction. Scanning the red-and-green-lit Christmas tree hull opening indicator panel, the chief reported, "Green board; pressure in the boat." The main ballast tank vents popped open with a *whoosh!* as the chief, his hands a blur, pulled the hydraulic vent valve control levers open one after another. Seawater flooding her ballast tanks, the *Spadefish* started to submerge. A landlubber might have thought he was witnessing pure chaos but in fact it was a series of practiced evolutions required to submerge the *Spadefish* in thirty seconds.

Underwood, looking down into the control room through the open hatch in the conning tower's deck, ordered, "Make your depth six-five feet."

The diving officer confirmed the order. "Six-five feet, aye." Lockwood, the deck under his feet angling down, held on and leaned away from the dive. He was in his element, among fellow submariners aboard a fleet sub.

The two men seated at the diving stand in the control room muscled the big nickel-plated wheels controlling the *Spadefish*'s bow and stern planes. As the *Spadefish* approached sixty-five feet—periscope depth—the diving officer asked for two-thirds speed, then ordered, "Blow negative to the mark!" Dewatering negative tank restored the sub's neutral buoyancy and helped trim the ship for submerged operations. Then, "Ease your bubble."

Underwood motioned for the periscope. The quartermaster yanked the hydraulic control lever in the overhead. The scope hummed out of its well, jerked to a stop at its upper limit. Underwood made a quick

360-degree scan. Satisfied that the area was clear of ships, he turned the scope to the marker buoys dead ahead, where the minelayers had sown a dummy field. "Coming up on the mine plant."

Satisfied that he'd demonstrated for his three-star boss how tight a ship he ran, Underwood ordered, "All ahead one-third. Stand by on FM sonar."

Lockwood hovered near the PPI scope, watching over the operator's shoulder. He saw a circular glass screen with concentric rings, like ripples from a stone dropped into a pond. The rings indicated ranges from one hundred yards to several thousand yards. Bearing lines, like the spokes of a wagon wheel, extended across the scope's face to all points of the compass. The circles and spokes allowed an operator to plot the position of any mine contacts relative to the position of the *Spadefish*. A thin, luminous line swept continuously around the circumference of the scope: If the sonar beam made contact with a mine, the contact would blossom into a bright green pear every time the line swept past its position.

No sooner had the *Spadefish* dived than Lockwood, elbow-to-elbow with Underwood, heard the sound of a bell, clear and undistorted, tolling from the FM sonar's speaker—the infamous hell's bells! The instant the bell began ringing, the sonar technician at the PPI console sang out, "Mine contact!"

Ahead and to the right of the advancing *Spadefish* danced a dummy mine on a deep-anchored cable.

Triumph! thought Lockwood. The damned thing actually worked! Not only did it locate mines, but with proper handling and a studied touch of its controls it could point a sub toward an open path through them. Even now the *Spadefish*, advancing submerged at three knots, swept on through the minefield, "blobs glowing on the sonar screen not so much like pears but like big juicy lightning bugs."[3] Lockwood was ecstatic. He'd put his reputation on the line to prove that FM sonar would work. And it did. As he slapped a still skeptical Underwood on the back, he saw with his own eyes that it was the secret weapon he'd dreamed of that would get his subs back into the Sea of Japan to finish off the Japanese.

Sure, FM sonar was only as good as the men who operated it. Both the operators and the captains of the sonar-equipped subs would need specialized training on the equipment, not just to learn how to take full advantage of its abilities and to overcome its limitations, but also to gain confidence that its employment against enemy minefields would open new horizons for the sub force. Until that moment off Oahu, when hell's bells and big, juicy lightning bugs announced the presence of mines, Lockwood, like every submariner, had accepted the fact that mines embodied an occupational hazard and that nothing could be done about it. Now, FM sonar would change everything. As operators gained experience and as the sonar underwent improvements, submariners could then embark in a great fleet of FM sonar–equipped submarines with little to fear from the great mine menace. Then and there Lockwood vowed that he would see to it personally that both captains and sonar operators received the training they would need on FM sonar before setting out on patrols.

To realize this goal, Lockwood would have to negotiate bureaucratic roadblocks and manufacturing bottlenecks. He also faced another serious problem over which he had no control. Though a prototype unit from UCDWR's labs had found its way aboard the *Spadefish*, future production units were destined not for the sub force but the Navy's minesweeper force. This despite the fact that the mine force had flatly rejected the units because their poor performance made minesweeping more dangerous than it already was using traditional methods of paravanes and cable cutters.

Fuming that the needs of the sub force had been ignored, Lockwood presented his argument directly to Admiral Nimitz for the acquisition of the unwanted minesweeper FM sonar units for submarine use. Meeting with the fleet admiral in mid-July, Lockwood made a presentation that included details of the tests conducted in San Diego and off Oahu in the *Spadefish*. While he was at it he made a pitch for getting his hands on as many sets as he could to equip submarines undergoing overhaul on the West Coast and, if possible, some of those based in Pearl. As Lockwood talked and talked, Nimitz, his pale blue eyes assessing Lockwood's effervescent enthusiasm for his pet project, listened carefully. Lockwood completed his presentation with, "Eventually, Admiral, with this new sonar, we'll crack the Sea of Japan without losing a ship or man to the minefields."[4]

Nimitz was a realist. Though he shared Lockwood's enthusiasm for such a mission and appreciated that ComSubPac was not a man to unnecessarily risk men's lives on harebrained schemes, he may have had doubts that a force of submarines equipped with Lockwood's magic gear could run the gauntlet without incurring losses. He may have even thought that Lockwood's chances of twisting arms at the Navy Department, which would have the final say on the diversion of sonar sets, were slim to none, and that nothing more would come of the mission. Nimitz, weighing the risks and potential rewards in Lockwood's proposal, gave it his approval. Even with Nimitz's approval in hand, Lockwood faced a tough battle. He would have to convince the Navy's bureaucracy in Washington that if his submarines were equipped with the FM sonar units destined for minesweepers, they would become one of the keys to ending the war in the Pacific sooner rather than later. To win that battle Lockwood would have to enlist the Navy's top admiral, CNO Ernest King, to his cause. The problem with this line of attack was that King in Washington and Lockwood in Pearl Harbor were based almost five thousand miles apart. A letter to King outlining the problem was impractical, and a lengthy telephone conversation over already jammed phone lines to the States meant solely for priority communications wasn't possible. How to get himself in front of King, even for a few minutes? Lockwood pondered this problem for a time until a solution presented itself via a secret communication to all commands in Hawaii: A top-level strategy meeting between President Roosevelt, General Douglas MacArthur, Admiral King, and Admiral Nimitz would take place in Honolulu in late July.[5] Lockwood saw his opportunity and seized it.

Admiral King arrived in Pearl Harbor days ahead of President Roosevelt, who had embarked from the States aboard the heavy cruiser USS *Baltimore* (CA-68). Fresh from a tour conducted by Lockwood of a submarine about to sail on a war patrol, King agreed to sit down with Nimitz and

Lockwood to discuss the FM sonar issue. Strange as it seems, King did not tour the *Spadefish* preparing to depart on her first war patrol from Pearl Harbor on July 23. The reason might be that her follow-up trials had not been as successful as the first one Lockwood had witnessed and that he didn't want to reveal that fact to King or Nimitz. As it turned out the *Spadefish* sailed with her FM sonar unit out of commission.

King quickly grasped the importance of Lockwood's proposal. He, like Nimitz, believed submarines could play a decisive role in the final collapse of Japan—they'd already brought Japan to the brink—and so he was eager to make it happen. The CNO, with a glance at a smiling Nimitz, nodded his assent to Lockwood's plan. Now all it would take to start untangling the red tape that had bound the FM sonar sets over to the mine force and divert them to ComSubPac's use would be a few words from King to his chief of plans. Little wonder that Lockwood felt he'd successfully moved heaven and earth to get what he needed to begin serious planning for a raid into the Sea of Japan. All he needed to make that happen were more submarines and more FM sonar sets. And one more thing, too: He needed submariners to buy into his plan.

CHAPTER SIX
Wolf Pack

Fresh from a two-week stay at the camps outside Fremantle set aside for the rest and recuperation of submariners, the *Bonefish*'s crew[*] reported aboard their ship, which had been refitted by the submarine repair unit of the tender USS *Griffin* (AS-13).

On September 5, 1944, the *Bonefish*, setting out on her sixth war patrol, stopped at the forward supply base established at Port Darwin on the north coast of Australia to top off with food and fuel. In addition to his orders governing the *Bonefish*'s patrol, Edge had orders to operate with the submarines *Flasher* (SS-249) and *Lapon* (that veteran of an earlier patrol in the Sea of Japan) as part of a coordinated attack group, in other words, a wolf pack. The pack's operating area encompassed the waters of the Sibuyan Sea south of Luzon, Philippines, and the South China Sea west of Luzon.

The German *Ubootwaffe* had successfully employed U-boat wolf packs against Allied convoys in the North Atlantic. These wolf packs had come perilously close to severing the only supply lifeline England had. It wasn't until the spring of 1943 that the Allies, after breaking the German Enigma submarine codes, began sinking hundreds of U-boats, thus saving England from almost certain defeat by Nazi Germany.

Unlike the Germans with their vast fleet of submarines, ComSubPac, hampered early on by having too few submarines to form effective wolf

---

[*]Ten to twenty percent of a submarine's crew rotated ashore between patrols. They were replaced by men from a pool of submariners who had sat out a patrol while assigned to submarine repair units aboard tenders.

packs, relied instead on freedom of action by individual subs for success against Japanese convoys. Unlike Allied convoys, Japanese convoys were smaller and consequently easier for a single submarine to attack. Whereas Allied convoys often consisted of eighty or more ships, the Japanese convoy commanders, lacking sufficient escorts, formed convoys of six to eight ships and sometimes as few as two. American subs didn't often encounter large convoys of fifteen to twenty *marus* and when they did, they usually had insufficient torpedoes to deal with such a large group of ships on their own. An American sub skipper in this situation had no choice but to pick out one or two ships to attack, knowing full well that as soon as the first torpedo went off the convoy would scatter all over the ocean, effectively ending any chance the submarine might have had to attack the other ships. Enter the wolf pack.

In the Pacific, two or three submarines operating together would attack by hitting the main convoy body from different locations, then, after pulling back, pick off the stragglers one by one. How effective U.S. wolf packs were was never fully decided. One problem was that U.S. wolf packs were too often hampered by poor voice radio communications between subs. Lockwood and Voge struggled to improve communications so as to better coordinate wolf pack attacks and, incidentally, to make sure the subs didn't torpedo one another. Submarine command organized about 120 wolf packs consisting of between three and seven submarines. In the main their operations succeeded, sinking more than a hundred Japanese ships. To prove that wolf-packing worked, early on the USS *Steelhead* (SS-280) and the USS *Parche* (SS-384) attacked a fifteen-ship convoy, sank five ships, and damaged many more. Yet as impressive as this action was, the USS *Rasher* (SS-269), acting alone, tore into a twenty-ship convoy in the South China Sea off the Philippines, sank four ships—including the twenty-thousand-ton escort carrier shepherding the convoy—and badly damaged four more. The *Rasher*'s amazing feat proved once again just how potent a single submarine captained by an aggressive skipper could be.

The Germans often utilized as many as twenty U-boats in wolf packs that relied on an unwieldy but effective shore-based communications net to coordinate their attacks on Allied convoys. The Germans sank a lot of ships using this method but lost a lot of U-boats and men. After all, a pack of twenty U-boats swirling like wolves around a convoy in the North Atlantic presented a juicy target for Allied antisubmarine forces. The lesson ComSubPac learned from the wholesale destruction of German U-boats was that small wolf packs made for a more compact and survivable weapon able to nibble away at Japanese shipping despite the ever more effective antisubmarine measures the enemy employed as the war progressed.

The _Bonefish_ arrived in her patrol area on September 21. Patrolling the Tablas Strait south of Luzon, she once again encountered dozens of sailboats and small craft, mostly local fishing vessels and luggers. They were far too numerous to board and inspect either for cargoes destined for the Japanese or for submarine-spotter radios. After spending several unproductive days in this area Edge moved westward into Verde Island Passage. So far, other than sailboats, the area, reputedly teeming with Japanese freighters, looked pretty desolate.

Past midnight on September 27, Edge received orders from Fremantle to join the _Flasher_ and _Lapon_ off Luzon for a wolf-pack operation two days hence. Both subs had reported sighting targets off the southwestern Philippine coast, a major north-south Japanese shipping route. Fremantle had orchestrated the wolf pack's moves based on intelligence gleaned from decrypted Japanese radio messages that included valuable information on the routes convoys tended to use and the makeup of the convoys themselves and their cargoes.

Based on this information Edge shaped a course through the Verde Island Passage north of Mindoro into the South China Sea, and on four engines headed for the area he'd been assigned. Entering the South China Sea, _radar contact!_

Edge altered course to begin tracking the target. Two hours of hard running at flank speed brought the _Bonefish_ into torpedo-firing range of a large zigzagging northbound ship and two echo-ranging escorts. Fighting the clock—dawn's arrival and, with it, air patrols out of Manila—Edge, ready to launch an attack, tucked in behind the ship and her escorts. In position downwind of them, Edge smelled fuel oil, a tip-off that he had a loaded tanker in his sights.

Sidestepping both escorts Edge made his run-in. He fired the bow tubes at the target's broadside, then spun the *Bonefish* on her heel to fire the stern tubes. Before he could work a new setup, an explosion erupted near the target's stern. An instant later a blinding flash and a huge ball of flame topped by boiling black smoke rolled into a sky now tinged with dawn. The flash and ball of flame momentarily turned night into day and made the *Bonefish* and the Japanese escorts stand out like yachts at a regatta.* Scorching heat from the fireball whipped over the *Bonefish*. To the men on her bridge it felt like a dragon breathing fire on them. For Edge, his vow to make amends for his failure to sink that ship off the coast of Zamboanga during the *Bonefish*'s last patrol had been fulfilled.

With morning light spreading fast in the east and the sure arrival of planes imminent, Edge didn't hang around to inspect his handiwork. He rang up flank speed for the wolf pack area and retired westward, away from the smear of black smoke lying low on the sea.

Early the next day a radar contact off Mindoro developed into another oiler and two escorts, one of them a sleek destroyer. With her low freeboard and hidden Plimsoll mark the oiler looked fully loaded.

Once again Edge worked in and fired six torpedoes. This time torpedo gremlins attacked in force but couldn't save the doomed target. One of the torpedoes tore off on an erratic course; three others missed the target completely. Two evaded the gremlins and crashed into the ship. One of them hit just aft the MOT—middle of the target—sending hot debris and sparks pinwheeling skyward. The other one hit just forward of the MOT with the same effect. Hull ripped open, oil gushing from ruptured bunkers, the tanker† vanished from the *Bonefish*'s radar screen.

---

*Despite all the explosions and the pall of smoke from the ship as she settled in the water, JANAC claimed that she had not been sunk, only damaged. Nevertheless, Edge noted in his patrol report that she'd disappeared off radar, a sure sign that she'd been sunk.

†JANAC's postwar accounting confirmed the sinking of the two-thousand-ton *Anjo Maru*.

Three days later Edge made contact with the *Flasher*-led wolf pack. After an exchange of information, including details of his attacks on the two tankers, Edge began patrolling off the west coast of Luzon, where he found plenty of targets.

Convoys hugging the Philippine coast, some of them containing as many as ten ships, lit up the *Bonefish*'s radarscope. But Edge was driven off by echo-ranging escorts, which, alerted to the presence of submarines, intermittently dropped patterns of warning depth charges. Planes patrolling out ahead of the convoys dropped bombs whenever their pilots thought they had a submarine in their sights. In addition to the enemy's antisubmarine efforts, a persistent and heavy overcast interfered with celestial navigation. Without accurate position fixing, conning a submarine in and around coastal convoy routes trapped with shoals, reefs, rocks, and bars was pure guesswork and also dangerous.

It wasn't until October 10 that Edge's luck changed and the weather moderated. Patrolling submerged off Cape Bolinao, the westernmost point of land midway up the coast of Luzon, the sound watch reported echo ranging from escorts. A careful search by periscope revealed thin columns of smoke from ships hull-down over the horizon. As the smokers plodded over the horizon toward the *Bonefish*, Edge got a good look at a ragged eight-ship convoy protected by four echo-ranging escorts and a low-flying plane. "Battle stations torpedo!"

Men leaped to their stations as Edge started an approach. Though the *Bonefish* wasn't far off the convoy's base course, the approach was hampered by heavy swells that made depth control difficult and necessitated high and prolonged periscope exposure to keep track of the convoy. If the escorts were alert they'd spot the raised periscope and its white feather streaking through the sea. But Edge didn't worry about that or the plane buzzing over the convoy like a mayfly. Should the pilot spot the *Bonefish*'s dark hull moving underwater in the direction of the convoy, Edge would go deep.

He worked the *Bonefish* between the two columns of ships. He did it confident that the pilot of that plane was napping or, more likely, that the ruffled, leaden sea had camouflaged the *Bonefish* sufficiently to allow her to get close enough to count nine freighters and a lone tanker. With an eye on the nearest escort, Edge picked out a fat, juicy target and . . . All of a sudden he saw an escort change position from starboard to port, crossing the *Bonefish*'s bow less than a thousand yards away! Edge also saw the escort two-block a flag, a warning to the convoy that it had a sub contact! Edge ignored him for now; he had that fat, juicy target at the end of a long, two-thousand-yard torpedo run. *Fire One!*

> **1108-25 Commenced firing bow tubes at medium AK [cargo ship]**
>
> **1108-41 Checked fire after fourth one . . . A hasty and eye-filling glance at the escort showed him slightly on our port bow with a bone [in his teeth] looking as big as he was . . . and headed to pass something less than 100 yards down our port side. Could practically see their lookouts staring down the scope at us.**
>
> **1109-22 Resumed fire with the [remaining] two [torpedoes in the forward nest] . . .**
>
> **1109-35 Good, loud unmistakable torpedo explosion, timed to be the first torpedo hitting the target.[1]**

Two more hits! The familiar sound of exploding warheads rumbled through the *Bonefish*. Three minutes later Edge and his crew were

> **. . . hauling ourselves down to 350 feet fast when the first depth charges started going off. During the next twelve minutes, 35 depth charges exploded, none of which seemed very far away, and some of which could hardly have been closer without causing serious damage—or so it seemed! [Electrical] cables were pushed in as much as four inches; every compartment had a quota of shattered light bulbs; fuel tank inboard vents were [forced] open, getting considerable fuel oil**

**over After Battery and Forward Torpedo room
decks. . . . But nothing serious turned up, and we felt
pretty proud of our [strong] hull.**

**1125 Settled down to evading at 4 knots, pulling away
from the coast to westward. Two [escorts] seemed to
stay back . . . (picking up survivors?).[2]**

Forced down by the depth charging, Edge didn't know what had happened topside, yet was reasonably certain that if they'd not sunk a ship,
several had been damaged.[*] When it was all over Edge made contact with
the *Lapon* and *Flasher* to report the convoy's position. With luck the wolf
pack might be able to intercept what remained of it.

The next day: "Captain, echo ranging."

The periscope lurched out of its well. Edge snapped down the training
handles, put his eye to the ocular. The exec stood on the opposite side of
the scope, his hands on Edge's gripping the training handles, and rotated
the scope onto the reported bearing of the echo ranging.

Once again Edge saw smokers, six or seven of them. Soon enough the
masts of hull-down ships heaved over the horizon just as those ships had
yesterday. Fifteen minutes later ten ships and their escorts were strung out
in a long southbound column hugging the twenty-fathom curve off the
coast near Santa Cruz. It seemed likely that the convoy was headed for an
anchorage there or at Manila. The trouble was, the submerged *Bonefish*
was so far off the convoy's base course she couldn't catch up to its main
body. Because it was still daylight the only thing Edge could do was to
intercept the tail end of the convoy as it steamed down the coast. His
scheme worked until the convoy seemingly evaporated into thin air. "Lost
contact," Edge reported. A taxing periscope search turned up nothing.

Three blasts of the Klaxon reverberated through the ship. Daylight
or not, Edge was on the surface with four straining diesels to find the

---

[*]JANAC confirmed damage to two, possibly three ships.

convoy and send a radio message to the *Flasher*: "Convoy. Chasing. Edge sends."

Despite intermittent radar contact with the convoy, it somehow slipped away. Peppered with small bays and shallow bights, the Philippine coast offered suitable refuge to small convoys but not ones made up of ten ships and escorts. A convoy that big would need a body of water almost the size of Subic Bay in which to anchor. Acting on that fact the *Bonefish* reached Subic Bay around midnight to discover that the convoy was nowhere in sight. At that moment Edge realized that by racing south down the coast he'd gotten ahead of it and that if he reversed course he'd run into it.

While Edge hunted for the elusive convoy, ComSubSoWesPac was busy handing out new assignments to patrolling submarines: A radio message from Fremantle directed the *Bonefish* and *Lapon* to assume lifeguard stations off Cape Bolinao on October 16.

Lifeguarding—plucking downed U.S. fliers from the sea under the noses of the Japanese—was an important service submarines had been providing to the Army and Navy air forces. With the advent of air strikes on Japan launched from the Marianas, lifeguarding rapidly expanded from a sideline for patrolling submarines to a nearly full-time job. By war's end U.S. submarines had rescued 504 airmen ditched at sea. In the Philippines, the Navy had begun launching air strikes against the Japanese in preparation for the invasion of Leyte, scheduled for October 21. Lifeguarding submarines would be needed for the rescue of downed Navy pilots flying air-support missions during that operation.

As he patrolled back to the north, Edge expected to find the convoy standing down the coast. The *Bonefish* pounded along, radar sweeping, lookouts stationed in the periscope shears with binoculars hunting for the white bow waves and jagged black profiles of their prey. Hours passed. The radar men peaked the SJ: Nothing out there, they reported, but the Philippine coast and a vast, empty sea. Weary lookouts just shrugged coming off watch: Nothing in sight but water, they said. Time passed and with it the prospect of a swift, hard-hitting attack on an unsuspecting row of targets slowly evaporated. Edge, convinced that the convoy had some-how slipped past them or had anchored somewhere out of sight, finally abandoned the search. The *Bonefish* continued north to patrol off Cape Bolinao lighthouse, which was not far from the area where she had been assigned lifeguard duties.

Edge patrolled the area submerged, searching for targets. He didn't have long to wait. "Captain to the conning tower!" The periscope watch reported a thin column of smoke on the horizon. Edge eased behind the scope and saw a zigzagging, smoking freighter. Minutes later a second smoking ship heaved into view behind the first one. Edge put the two 2,000-ton freighters in the crosshairs of the attack scope, then swung into action.

Four torpedoes aimed at the trailing ship streaked from the sub's stern tubes. A heavy explosion erupted on the target and Edge watched her go down under a thick cloud of smoke and steam. The lead ship, which he thought was another freighter, turned out to be an escort-type vessel armed with depth charges. As he watched the hapless escort from a safe distance through the periscope, Edge noted Japanese sailors struggling to roll heavy ash cans over her sides.

1140 . . . nothing visible [Edge reported] but the nearby land and one maru, definitely the one which had been leading, now dropping [depth charges].

1145 Went back to 250 feet to evade at 4 knots.

1156 Breaking-up noises became loud enough to be heard through the hull for the next forty-five minutes, these noises were a little like small underwater explosions. . . .

Between the thundering of depth charges the submariners heard the crash of collapsing bulkheads and internal compartments, the screech and groan of cargo tearing loose, the thousand other popping and crack-ling noises that a ship makes when it breaks up and sinks.

1226 Back to periscope depth, all clear except land and masts of escort maru.[3]*

———

Concluding her torpedo work off Luzon, the *Bonefish* arrived at her assigned lifeguard area the morning of October 16. Edge began patrolling a five-mile-wide figure eight, waiting for a call to pick up downed fliers. It was boring but important work, submariners told themselves. But after the thrill of a chase punctuated by the staccato drum of diesels at full song, booming torpedoes, cracking depth charges, and ships breaking up, playing acey-deucey and cribbage and watching the same old scratchy movies over and over again could almost make a sub sailor long for shore duty. Submerging only to dodge enemy planes and patrol boats, the *Bonefish* lazily churned her figure eight. What little action there was centered on the radio room, where the radiomen on watch

> . . . carefully guarded the assigned VHF primary and secondary HF channels [Edge wrote] . . . but the only indications we ever had that an air strike or strikes may have been in progress were that on the 17th and 18th, radio MANILA (commercial) went off the air part of the day with an air raid warning. In the evening of both days, also, short meaningless snatches of a few American phrases were picked up on the primary HF channel. All planes sighted closely enough for recognition had appeared to be enemy types (float planes, etc.).[4]

On the eighteenth and with night coming on, the *Bonefish* received a radio message vectored through Fremantle reporting the position of two Navy airmen adrift in a rubber boat about eighty-five miles north of the *Bonefish*. The men had reportedly been in the water for nine hours. Edge didn't have to be told that if they weren't picked up soon they stood a good chance of being captured by the Japanese. Edge bent on four engines to get there as soon as possible. Even so he knew it wouldn't be easy to find

———

*JANAC confirmed the sinking of the 2,500-ton *Fushimi Maru*.

two men in an inflatable lifeboat tossing around on the South China Sea in the dark.

The steady thrum of four diesels signaled that this was serious business. For hours the *Bonefish* sped up the coast. Heading to the rescue she passed two burning ships in a harbor. Edge surmised that they'd been bombed by the airmen he was hunting for. Shortly after midnight, as the *Bonefish* approached the aircrew's reported position not far from the harbor with its burning ships, a red flare arced across the night sky. Americans or Japanese? There was no way to tell. It was too dark to see anything except black sky and black water.

Exercising caution Edge closed in, mindful that Japanese ships often used red flares to signal other ships at night. Even so, Edge gave the okay to fire an answering flare from the *Bonefish*'s Very pistol, which was answered immediately by another red flare. Still on guard, Edge eased the *Bonefish* toward the dying flare. The Navy fliers in their rubber boat saw the shadowy outline of a submarine approaching, heard her sputtering diesels, and fired another flare. This one brought the *Bonefish* within shouting distance of them, fifty yards off her starboard bow.

Happy *Bonefish* sailors hauled aboard the pilot and enlisted aviation gunner's mate of a Navy SB2C Helldiver from the USS *Bunker Hill* (CV-17). They'd been shot down during a skip-bombing attack on the burning ships Edge had spotted earlier. The Helldiver had caught a burst of antiaircraft fire from the ground and went down with a shot-up engine. The two airmen had rowed toward land all afternoon against strong currents that eventually pushed them back out to sea. With the men safely aboard the *Bonefish*, Edge turned her nose south for Australia.

Edge's sixth patrol report received glowing endorsements from Admiral Christie and from Edge's squadron and division commanders. They praised Edge for his excellent coverage of the patrol area, his aggressive attacks, and the rescue of two airmen. For this outstanding patrol Edge was recommended for the Navy Cross, the *Bonefish* for a fifth Navy Unit Commendation. Flush with success Edge and his crew received new orders via the commandant of the Twelfth Naval District, San Francisco, California:

```
PROCEED HUNTERS POINT LAND TORPEDOES X
AWAIT ARRIVAL OF STATION PILOT TO PROCEED
TO BETHLEHEM STEEL FOOT SIXTEENTH STREET
DOCK SAN FRANCISCO FOR OVERHAUL X CONSULT
ASSISTANT NAVAL INDUSTRIAL MANAGER SAN
FRANCISCO FERRY BUILDING RE WORK REQUIRED
X COMTWELVE SENDS⁵
```

The men were ecstatic. A return to the States meant extended shore leave to spend with family and friends. For Lawrence Edge it meant that he'd be reunited with his beloved Sarah and Boo, whom he'd yearned for so passionately during those long, lonely, grueling months at sea when all he could do was dream of a far-off day when he would return home to shower them with all the love he'd kept pent up in his heart.

Per standard procedure, Edge's sixth war patrol report worked its way through the Navy's chain of command, ending at the top with Cominch Admiral King. Edge's performance had caught Lockwood's eye back in Pearl Harbor and garnered ComSubPac's glowing endorsement. Lockwood directed his flag secretary to draft a private message for Edge's eyes only that would in time catch up to him in the States. Edge was a skipper to watch, a skipper who would be a perfect choice for the top-secret operation Lockwood had been planning and which he intended to spring on the Japanese before next summer was out.

CHAPTER SEVEN

The Long Road to Tokyo

Around the time that the *Bonefish* was standing out from Fremantle on her sixth war patrol in September 1944, Lockwood received the news he'd been hoping for: Eleven FM sonar sets assembled under a contract let to Western Electric by the Navy had been approved for shipment to ComSubPac. Lockwood was not only thrilled that his request for sets had been fulfilled so soon, but also that Admirals King and Nimitz had plowed past the Navy's nearly impenetrable wall of bureaucracy to ensure that the sets got into the hands of the sub force. Still, getting them installed in the boats was going to be a slow process, what with the production delays affecting electronic equipment destined for the fleet. FMS, as it was now being called, was no exception.

In late September and early October two FMS units arrived in Mare Island. One set was installed in the recently overhauled USS *Tinosa* (SS-283), a veteran of seven war patrols and skippered by her new CO, Commander Richard Latham, from New London, Connecticut. The other set went aboard the USS *Tunny* (SS-282), skippered by Commander George Pierce, who was born in Colón, Panama, and graduated from the academy in 1932. These two subs were the first to have retractable sonar transducers installed below their keels under the forward torpedo room.

Certain that delivery of FMS units to the sub force would accelerate, Lockwood began to lay plans for an important first step toward the mission that would eventually culminate in the raid on the Sea of Japan: detailing an FMS-equipped submarine to map the outer boundaries of minefields guarding that sea. A mapping operation might accomplish two things: Demonstrate under actual combat conditions FMS's ability to locate enemy mines, and produce the first-ever accurate plot of the extent and location of the outer reaches of the actual minefields that the subs would have to penetrate to reach their targets. Of course, one sub working alone could map only a small portion of the minefields, but it was a start. And as more FMS units became available other subs would follow this pioneering effort. Once an accurate map had been compiled it should then be possible to enter the sea by steering an FMS-plotted course through the minefields. The sooner Lockwood had the mine plots for study, the sooner he could prepare a mission prospectus for presentation to Nimitz and King.

As the FMS units in the *Tinosa* and *Tunny* underwent rigorous static testing by UCDWR technicians, their crews, like the *Spadefish*'s before them, received instruction in their use by Dr. Henderson himself. Tests aboard the *Tunny* off San Diego showed once again how effective the sonar system could be when it was properly tuned and in the hands of a trained operator who understood its peculiarities. Nonetheless, the *Tinosa*'s Richard Latham, who himself had serious doubts about the effectiveness of FMS, discovered that to a man, his crew shared his feelings. No one had to explain what was in store for them on a future war patrol. Worse yet, more than half of the men in Latham's crew wanted off the submarine. And who could blame them? Torpedoing Japanese ships was one thing; toying with live mines was something else. Rather than try to win them over, Latham simply had those men who wanted off transferred to other submarines. Latham would not be the only skipper to face this problem, one that would soon loom large in Lockwood's thinking as well.

Meanwhile, at Pearl Harbor, Lockwood grappled with a host of problems unrelated to FMS. Though submarines were enjoying greater success than they had heretofore (improved torpedo performance and bolder tactics accounted for a good part of this success), losses continued to mount. In four months half a score of subs and eight hundred men had been lost, most of them to direct enemy action but also, Lockwood suspected, to enemy mines. Then there was the USS *Tang* (SS-306). Operating off the coast of China, she had apparently been sunk by one of her own torpedoes that had made a circular run and come back to kill her. Whether there had been any survivors from the *Tang* was not known. At night, alone in his quarters, Lockwood grumbled, "God damn the torpedoes and God damn the Japs."

One bright spot, aside from FMS, was that with the conquest of Guam and Saipan, Lockwood now had forward operating bases thousands of miles closer to empire waters. He sent two submarine tenders ahead to refit and rearm submarines operating from those new bases. A floating dry dock would soon follow the tenders. Seabees had built a rest camp on Guam for submariners and were building one on Saipan. That Lockwood planned to set up his Submarine Pacific Advanced Headquarters on Guam in January was ample proof that the United States was rolling back the Japanese on every front in the Pacific. Admiral Nimitz planned to move his own headquarters there in January.

Nevertheless, FMS and the Sea of Japan raid were never far from Lockwood's thinking. In late November, the *Tinosa* arrived in Pearl Harbor fresh from Mare Island and FMS testing. Eager to see her perform, Lockwood put to sea with Latham on Thanksgiving Day for a run at the dummy minefield off Oahu. The results were generally good. It proved that the keel-mounted transducer in the *Tinosa* performed better than the deck-mounted one in the *Spadefish*. It easily detected mines, which the PPI and hell's bells rendered with exceptional clarity.

After completing an intensive course of FMS training under Lockwood's personal supervision, Latham and his men, half of them a draft of replacements, received orders to patrol the Nansei Shoto, Formosa, and East China Sea areas. It was Latham's first war patrol as a commanding officer and, it might be said, FMS guinea pig. The orders included a provision to explore the fringes of the patrol area for mines, especially around Okinawa, which was scheduled for invasion in early April 1945, and, if any were found, to plot their positions. This would be an important test of FMS, and Lockwood had a lot riding on its successful employment and on Latham's skill.

On December third Lockwood watched the *Tinosa* stand out from her berth at Ten-ten dock at the Pearl Harbor sub base. He believed that he'd done everything he could to ensure Latham's success. All he had to do now was wait for the results. After the *Tinosa* sailed, an optimistic Lockwood returned to his office to begin drafting a prospectus for the Japan Sea raid. This was it: The mission would either get the go-ahead or get the ax. The prospectus landed on Nimitz's desk a few days later.

TOP SECRET

From: The Commander Submarine Force, Pacific Fleet.
To: The Commander-in-Chief, United States Fleet.
Via: The Commander-in-Chief, U.S. Pacific Fleet.
Subject: Japan Sea—Patrol of.

1. At the beginning of the summer 1944 it was proposed to establish patrol in Japan Sea. The assumptions were that La Pérouse Strait was free from mines and that Tsugaru and Tsushima Straits were mined.

2. At this point a naval prisoner of war taken by U.S.S. BARB (SS-220) gave information, which JICPOA considers to be credible, that there are three rows of deep laid mines in La Pérouse Strait. The presence of such mine fields appears to be logical in view of the fact that the depth of water is shallow and that surface or near-surface mines would not be feasible on account of the ice during the winter months.

3. The project was, therefore, held in abeyance awaiting better information.

4. In the meantime, there has been developed a FM sonar which gives promise of being useful as a mine detector. The first installation of this equipment made on the SPADEFISH has not been too successful or reliable. A second set installed on the TINOSA and just recently tested

indicates that with good sound conditions mines can be detected up to a range of 600 yards.[1]

Lockwood explained that captured Japanese documents gave indications that the Tsushima Strait, though blocked by mines, presented the best possibilities for entry, though he didn't give details. Lockwood revisited the loss of the *Wahoo* when he described attempts by her and other subs to stanch the flow of traffic from Korean ports to the Japanese homeland. Finally, he outlined for Nimitz the steps ComSubPac had been taking to assemble reliable intelligence about the true extent of minefields. This included sending a spy to Vladivostok to obtain information regarding Japanese minefields from Russian shipmasters who had used the La Pérouse and Tsushima straits; aerial reconnaissance; exploration by submarine (the *Tinosa* mission); decrypts of Japanese radio transmissions regarding shipping routes in and around the areas in question; the taking of prisoners by subs operating in the vicinity of the straits; even the possibility of entering the Sea of Japan through the unmined Strait of Tartary during the ice-free summer months, if the Soviets would permit it, for a look-see. This last was clearly impossible: U.S.-Soviet relations were deteriorating by the day. Moscow would never allow an American warship, much less a submarine, to enter her territorial waters to probe or possibly attack Japanese shipping. It was their neutrality that allowed Soviet ships to transit the La Pérouse Strait unmolested. As it was, Lockwood would soon have to grapple with the problems posed by the possibility that Russian naval forces might begin operations in Japanese waters controlled by U.S. forces. But not now.

Nimitz gave Lockwood his unconditional support for the mission and urged him to collect additional intelligence to prove that access to the sea was possible and that there were ships in it worth sinking. The question of whether there were any ships left in the Sea of Japan worth the risks involved to sink them had nagged at Lockwood before and nagged at him now. The only way to know for sure how many there were was for his skippers to go in there and look around as Morton, Bass, and the others had done. Lockwood didn't doubt they'd find plenty of *marus* to sink.

He was thrilled to have his boss's endorsement because it meant he could count on Admiral King's as well. All he had to do now was get the FMS gear into the supply pipeline, install it on submarines, and train submariners how to use it. Notwithstanding Nimitz's approval, Lockwood the optimist knew what he was up against. Making the mission a reality required overcoming the technical problems plaguing FMS, described in a flurry of reports from Gaylord Harnwell, Malcolm Henderson, and ComSubPac's UCDWR liaison officer. Lockwood had confidence that the problems would soon be solved. He envisioned great things for his sub force, perhaps even for himself, and, of course, revenge for Morton and the *Wahoo*. Especially that.

As if to prove that success breeds success, Lockwood received word that the *Tinosa* had found and charted a minefield north of Formosa. For Lockwood, this first successful combat test of FMS proved that Emperor Hirohito's days were numbered and that the only thing left for him to ponder, besides his and his generals' fates, would be all the sunken ships littering the bottom of his private lake.

The report of the *Tinosa*'s success had reached ComSubPac via a radio message from Latham.[2] It caught up to Lockwood, commuting between Pearl Harbor and Guam to inspect his new digs. Use of the *Tinosa*'s FMS off Okinawa had resulted in bright green pears and clear, unambiguous bell tones indicating the presence of mines. Detection occurred at surprisingly long distances from the probing *Tinosa*, up to fourteen hundred yards in some cases. Further probing had resulted in questionable contacts and errant visual and audio signals. In some instances FMS had made contact with objects that may or may not have been mines. Latham realized that in this instance the lack of contacts didn't mean that there were no mines present, just that they'd not been found.

Working blind, as it were, the poor *Tinosa*'s crew had to endure hours of stress and dread. Latham saw the haunted looks on their sweat-burnished faces. He shared their crushing fear of being blown to bits while trapped inside the ship's hull. The work they were doing was not meant for cowards. Terrifying though it was, the mapping mission had to be completed; a lot was riding on it.

In all, the *Tinosa*'s reconnaissance proved a success. Even though the minefields that ICPOA believed had been planted so heavily around Okinawa never materialized, Latham found a few sparse fields—more than enough to satisfy him and his wary crew. Lockwood reviewed Latham's patrol report and pronounced it outstanding and a genuine breakthrough in the gadget's use against real mines, not dummies. Viewed from a purely technical perspective it proved that a submerged submarine equipped with FMS could plot mines with a surprising degree of accuracy. It also underscored how effective submarines could be when they utilized their inherent stealth and near invisibility to operate effectively against enemies in waters that were otherwise denied them. Stealth coupled to FMS would not only enhance the effectiveness of submarines against Japanese targets; it would make them virtually impervious to the enemy's defenses.

Elated by Latham's success, in late January 1945 Lockwood held the first of several operational planning meetings with his staff to review the intelligence developed by Latham. His report disclosed that though the mines plotted by the *Tinosa* had been sown in rows and in staggered depths, there was enough space for a submarine to squeeze through the gaps between them with room to spare. Of course, that was the layout of the plot around Okinawa, not the Tsushima Strait. Lockwood next wanted an accurate plot of the mines sown around the entrance and inner approaches to the Tsushima Strait. This data would then be used to create a map of the minefields for use by the raiding subs.

In his report Latham had also made note of the Kuroshio Current off Formosa and the force it had exerted on the *Tinosa*, especially when she was submerged. The current, he reported, was equally strong if not stronger around southern Kyushu than around Formosa. Its presence would definitely be felt in waters around the Tsushima Strait and would need to be taken into account for future mission planning. Lockwood wanted more data, and the only way to get it was to send more FMS-equipped submarines into mined areas around the straits targeted for entry.

There was another issue looming that Lockwood and his staff would soon have to address: It was becoming all too apparent that the skippers and crews training for operations with FMS had growing doubts about its effectiveness and, with it, their chances of surviving an encounter with a minefield. But first, Lockwood had to tackle FMS production delays in order to get more subs to sea so they could probe and map the area they'd be operating in and, in the bargain, prove to those doubting sub crews how reliable FMS was. Then, despite all the success Lockwood had enjoyed so far, he suddenly ran head-on into an FMS brick wall.

CHAPTER EIGHT

The Magic Loses Its Magic

While the scientists and technicians at UCDWR in San Diego grappled with the seemingly endless problems plaguing FMS, the *Bonefish* sailed under the Golden Gate Bridge on November 18, 1944, bound for Hunters Point in San Francisco Bay.

Lawrence Edge and his men were relieved to be home and excited by the prospect of seeing loved ones and friends. Lawrence took a part of his accumulated leave in Atlanta with Sarah, little Boo, and his parents. After the Christmas holidays the three of them returned to San Francisco for the remainder of Lawrence's stay at Hunters Point. The crews of submarines undergoing long overhauls usually brought their wives and children to San Francisco to spend as much time together as possible, living in rented apartments or in Navy housing. This arrangement gave Lawrence, Sarah, and Boo the opportunity to be together in one place for the duration of the overhaul. It didn't make up for the long separation that the war had imposed upon them, but for now, at least, their stay in San Francisco helped shorten the time until their lives would return to normal.

The *Bonefish* was due to complete her overhaul in mid-February. After postoverhaul shakedown and training, she was scheduled to depart San Francisco for Pearl Harbor in early March. To meet that schedule Lawrence put in ten- to twelve-hour days supervising repairs, handling paperwork, and dealing with personnel and their training schedules. The Bethlehem Steel Submarine Repair Basin at Hunters Point was a huge facility crowded with ships, dry docks, mobile cranes, and thousands of workers, both civilian and Navy. The *Bonefish* wasn't just overhauled; she was modernized, too. The work performed was extensive. It included major renovations and structural improvements to her hull, machinery, and interior compartments as well as heavier armament, updated radar, and a host of other improvements designed to enhance her combat efficiency and survivability. She also received an ice-cream maker and a Coca-Cola dispenser, welcome additions that would keep crew morale at its peak during long patrols.

Remodeled, reconditioned, thoroughly modernized, and in her fresh dark-gray-and-black camouflage paint scheme, the *Bonefish* was ready for operations in northern Pacific waters, where the submarine war now focused.

The confidential letter Lockwood had drafted earlier for Lawrence's eyes only caught up to him in San Francisco at the end of November. What he read must have pleased him immensely, for it placed him in the first rank of submarine skippers.

> *Dear Edge:*
>
> *The Admiral and Chief of Staff like to have the commanding officers know their reactions to war patrols. Inasmuch as they wish these reactions to be entirely "off the record," the Admiral has asked me to write letters to commanding officers regarding these comments. The comments may be of a praiseworthy nature or critical in type when deemed necessary. This letter is of course private and of an unofficial nature. Their comments made on your last patrol are as follows:*
>
> *"Very excellent patrol. Attacks well planned and executed.*
>
> *"I congratulate you on a good job. Have a good rest in the U.S. and come back with your chin out."*
>
> *C. A. [Lockwood], Jr.*
>
> *"An eventful patrol, chock o block with contacts, both surface and air."*
>
> *"You got a very nice bag and continued the fine reputation of the BONEFISH. Targets, available and not attacked, outlucked you."*

"The coming cruise to [West Coast] is primarily to put BONEFISH in an A-1 fighting condition; secondarily, for a deserved [rest]. Combine the two and return ready for Sea."

C. [Lockwood]

Please remember that this is a private affair between you and the Admiral and is meant to be helpful criticism, where present. No official record is kept of the above.

E. E. Yeomans [signed][1]

Lockwood believed that Lawrence would soon have an important role to play in future submarine operations, and in the submarine force itself.

Lawrence, Sarah, and Boo settled into their temporary quarters in a half Quonset hut at Hunters Point. Housing was at a premium because San Francisco's population had surged from a prewar high of 400,000 to over half a million, most of whom were employed in the shipyards in the surrounding area. This increase did not include the hordes of soldiers, sailors, and marines stationed at the Presidio, Treasure Island, and other bases.

Lawrence spent his days on the *Bonefish*, returning in the evening to his little family, clothes and skin smelling of diesel oil and hard work. Together, on one of the rare days he had off, and with a car borrowed from the submarine welfare motor pool, they went to the zoo or took a long drive up the coast. Sometimes on evenings and weekends, when a sitter could be found to care for Boo, Lawrence and Sarah went out to dinner or to a movie, alone or with other Navy couples. Around this time, too, Sarah learned that she was pregnant. The couple was thrilled. Lawrence wanted a boy, and though it was very early in Sarah's pregnancy he started calling the baby "Junior." The only thing that dampened Lawrence's and Sarah's excitement was the inexorable countdown to his departure and their separation.

By mid-January the *Bonefish* had undergone dockside testing and static dives. Crew training had become a daily ritual. Lawrence, fully engaged in this work, found time to write his parents that:

> We received a copy of a letter on the boat the other day which was from [Lockwood, Nimitz, and King] recommending that I be awarded the Navy Cross for our last patrol. If King's outfit ashore approves, I guess I'll eventually get the award— which naturally makes me feel pretty good, since that is next to the highest decoration the Navy gives (only the Medal of Honor is higher).

This was the first of three Navy Crosses won by Lawrence, all with effusive support from Lockwood.

Toward the end of the *Bonefish*'s stay at Hunters Point, a truck arrived dockside with a crated FMS unit. The crew had been asking questions about the strange modifications made to their ship's keel, and about the shaft protruding through the main deck above the forward torpedo room. Some but not all of these questions were answered when technicians from UCDWR arrived to supervise the installation of the FMS electronics stack and its associated components. As to its purpose, well, that was still a mystery.

It wasn't a mystery to Edge. An electronics expert, he understood the basic principles of FMS. And like every skipper who had had FMS installed in his ship, he may have had misgivings about the plans sub command had drawn up for its use in his ship. The same was likely true for the *Bonefish* sonar technicians who had been sent to San Diego for training on FMS, and upon their return to Hunters Point found a unit had been installed in their sub. With that and with UCDWR technicians swarming over the ship with test gear and schematics, it became all too clear that the *Bonefish* would be hunting mines—a chilling prospect for sailors accustomed to hunting Japanese ships.

As work on the *Bonefish* and the other submarines at Hunters Point progressed, Lockwood's war of all-out attrition against Japanese merchant shipping began drawing to a close. Targets worthy of torpedoes were proving hard to find. In 1944 alone, the sub force had sent more than six hundred ships, merchant and combatant, representing 2.7 million tons, to the bottom of the ocean. Now, in January 1945, sinkings had begun tailing off precipitously, a sure indication that the Japanese high-seas merchant marine was all but extinct. This development gave Lockwood more time to devote to his Japan Sea operation. Then, just as things began to look their most promising, Lockwood hit that FM sonar wall. It arose when another FMS submarine, the USS *Bowfin* (SS-287), arrived in Pearl days ahead of the *Tunny*, which had been undergoing modifications at Mare Island. The *Bowfin*'s CO, Commander Alexander "Alec" K. Tyree, a native of Danville, Virginia, and a 1935 academy graduate, was eager to get under way on a war patrol.

A test of the *Bowfin*'s FMS in that dummy minefield off Oahu showed that it could easily detect mines at close range but not beyond two hundred yards, a performance that Lockwood considered unacceptable. His minimum requirement was six or seven hundred yards; the *Tinosa*'s and *Spadefish*'s units had turned in far better performance. Nothing that the technicians did made the *Bowfin*'s unit recover its sensitivity, and until it did, Tyree would have to delay departure on patrol.

Lockwood didn't witness these tests firsthand; he was at his new headquarters in Guam. And though he was a patient man the clock was ticking down on his pet project. He wrote to the project managers at UCDWR and urged them to speed up production so that more sonar units would become available to outfit more subs and to replace the ones that had problems. He also urged Rear Admiral J. A. Furer, the officer in Washington coordinating the Navy's research and development group, to put some muscle behind his efforts on behalf of the sub force to open up the production pipeline.

Another complication dogging Lockwood was the announcement of a competition to be held between several different sonar detection systems, arranged by officers in charge of Navy research and development who believed that there were better mine-detection systems out there than FMS. The three competing units had an alphabet soup of names, such as MATD (Mine and Torpedo Detector), SOD (Small Object Detector), OL

(Object Locator), and STU, a device under development by the British Admiralty. Lockwood fumed over the Navy's penchant for bureaucratic interference in a field in which deskbound sailors had no expertise. He vented his frustrations in a letter to his friend Rear Admiral Charles W. Styer, then ComSubLant. Styer had connections on the East Coast, and Lockwood hoped to use Styer's influence to head off the competition, which Lockwood believed was a waste of time and had come too late in the game. But because UCDWR's FMS still wasn't good enough for Lockwood's purposes, he was willing to consider any other type that had promise, so long as its manufacturer could meet ComSubPac's production and mission schedule.

Lockwood noted in a letter to another friend, Captain Frank C. Watkins,[2] head of the submarine desk at Main Navy, that there had been so many of these projects that he couldn't keep them all straight. "We continue to hear about ER Sonar which you say is 'not in this war.' I want something which will take me through Tsushima Straits yesterday—and I don't want to send a boat through there without a mine detector. We can lose boats fast enough without doing that."

In another letter to Styer, Lockwood reported, "The third FM Sonar came in on Bowfin, and while the mine detection was very excellent, the range was only two or three hundred yards. . . . Fourth set arrived on Tunny this week, and I will see what she can do. . . . I know that various bureau and east coast experts say FM Sonar is no good, but after all it is the best we have got and I would appreciate very much if these adverse experts could get the lead out of their pants and produce something better. We don't care what it is called provided it does the job."[3]

Lockwood expressed his concern to Furer that diminishing resources allocated by the Navy for the development of technologies designed to enhance the effectiveness and survivability of submarines would hamper their missions, which had become more dangerous than ever, given the growing effectiveness of Japanese antisubmarine countermeasures. Lockwood topped his priority list of needed equipment with FM sonar and, possibly to impress Furer with the urgency of the matter, divulged his plan to use it to penetrate the Tsushima Strait into the Sea of Japan.

Though not fully satisfied with the *Bowfin*'s FMS performance, Lockwood sent her on patrol but not into mined waters. He wanted Tyree to have a crack at using FMS, if not to detect mines, then to detect Japanese patrol boats to see if FMS could work in conjunction with "cuties," the new Mk 27 electric homing torpedoes slowly coming into service, several of which had been put aboard the *Bowfin* for testing on real targets. Apparently BuOrd had learned a lesson from the Mk 14 fiasco. The USS *Sea Owl* (SS-405) had recently nailed two patrol boats with cuties, a performance that had convinced Lockwood that more were needed and fast. He may have hoped that if Tyree could improve on the *Sea Owl*'s score, more cuties would enter production.

In fact Tyree didn't use his FM sonar at all and didn't fire any cuties during the *Bowfin*'s fifty-seven-day patrol. Using Mk 14s and Mk 18s he sank only a destroyer. Tyree went on to exchange gunfire with two antisubmarine picket boats and rescue two downed Navy fliers. That Tyree didn't use his cuties had nettled Lockwood. Typically he refrained from criticizing skippers for actions taken or not taken during war patrols; he understood that circumstances dictated what actions a CO took, and Lockwood didn't like to second-guess his skippers. Even so it seemed a waste of an opportunity to acquire more data on torpedo performance.

Meanwhile another setback arrived with the *Tunny* from Mare Island. Her FMS was only marginally better than the *Bowfin*'s. The unit could detect mines out to 350 to 400 yards, but it had a mysterious blind spot beyond the sub's bow.

Lockwood wrote Watkins, "As you will remember, I have been trying to boost this FM Sonar since December 1943, when I saw the first model at San Diego. I thought it held promise as a mine detector and hoped for big improvements with succeeding installations, but I must say that Bowfin and Tunny have been a sad blow to me. . . . Sorry my last two letters have been sobs, but sometimes I get a bit discouraged."[4] Watkins wrote back that everything that could be done to speed up debugging and production of FMS was being done.

As Lockwood's entreaties began to move through the Navy's bureaucracy he took matters into his own hands to find solutions to the myriad problems afflicting FMS. Typically it had always been up to submariners to solve the problems that bedeviled the force—diesel engines and faulty torpedoes, for instance—and this time it wasn't any different. Lockwood and his submariners would have to work the bugs out of FMS that hadn't been worked out at UCDWR. To do this he initiated a robust training program that went beyond the training program at UCDWR: specifically, how to use FMS but also how to tune and properly maintain it aboard ship. Somehow the sub force would have to find the skilled electronics technicians needed for this work, even if Lockwood had to shanghai them from the surface Navy into the submarine Navy.

Released from FMS training at Pearl, the *Tunny* sailed for Saipan, where she was met by Malcolm Henderson and two technicians from UCDWR. At Lockwood's request the men, aware of Lockwood's personal training initiative and its growing importance, had flown on ahead to meet the *Tunny* to try to straighten out her FMS electronics. Lockwood had shifted FMS training to Guam, where he could take an active hand. He not only went to sea aboard the *Tunny* with Henderson and the UCDWR technicians; he also operated the gear himself, twiddling the knobs and dials on the PPI in the conning tower as she eased through dummy minefields off Guam.

His presence and hands-on approach instilled confidence in the men that the mine detector wasn't their enemy but their protector. He explained that it would allow them to operate safely in formerly restricted areas in the course of their regular war patrols. Lockwood, always the big-picture man, knew that as U.S. forces moved closer and closer to Japan, force commanders would need accurate information on the location of minefields in waters stretching from southern Japan to Formosa in which invasion fleets would have to operate. Who else but his submariners with their magic mine gear would be capable of mapping those fields? Lockwood also understood that this work that the force might be called upon to provide could possibly scuttle his plans for the Japan Sea raid. He had to resolve the problems inherent in FMS so he could launch the mission as soon as possible, before his subs were relegated to a minor supporting role while Admiral William F. Halsey's Third Fleet and Admiral Raymond A. Spruance's Fifth Fleet crushed the Japanese and got all the credit.

Lockwood was under no illusions about how hard it was going to be to sell FMS. The difficulty he faced was amply demonstrated by the incredibly poor performance turned in by the *Tunny*'s FMS, primarily its inability to consistently detect mines and, when they were detected, its feeble visual display and strangled bell tones. The *Tunny*'s crew were not only restless and unsettled by what they saw, but, within earshot of Lockwood, they openly voiced their feelings of mistrust and doubt, even questioning the sanity of officers who would fob off such a device on the sub force. Even Pierce, the *Tunny*'s CO, had his doubts, though he was willing to suspend final judgment until Lockwood and Henderson had had a crack at improving the gear's performance. In all it was a failure that might have defeated even the most ardent supporter of FMS technology. Yet Lockwood's enthusiasm for its potential never flagged, not in public and never at sea. Not even when faced with the seemingly insoluble problems that bedeviled even Harnwell and his experts, who often worked until dawn tearing apart the gadget's innards to diagnose why its tubes and circuits overheated, shorted out, or just plain refused to operate.

Lockwood's faith in FMS, as well as Henderson's and his technicians' dedication, finally paid off when another test aboard the *Tunny* got the results Lockwood knew it was capable of. Now, instead of seeing mushy greenish blobs and hearing muffled tolling, Lockwood and Henderson saw luminous green pears and those icy, clear bell tones submariners loved to hate. And with those improved results there was a change of attitude by the *Tunny*'s crew. Maybe, just maybe, this goddamned thing would work after all!

Before the *Tunny* departed on her next war patrol, shipfitters installed a set of clearing lines around her many hull appendages and topside fixtures—diving planes, cleats, stanchions, etc. Made from lengths of thick cable welded to the submarine's hull, the lines were there to prevent mine cables from snagging and being dragged down into contact with the sub and detonating. It was a crude fix that would prove troublesome later when it was least expected.

Confident that the *Tunny* was as ready as she could be, Lockwood gave

Pierce sealed orders to find, plot, and penetrate a minefield fringing the East China Sea near Kyushu. Could Pierce do it? Lockwood believed he could, and so did Pierce. At some point the attempt had to be made if the invasion of the Sea of Japan were to get off the ground. Lockwood was so confident that he requested permission from Admiral Nimitz to accompany Pierce to prove how much faith he had in FMS and to see it in action against the real thing. He reasoned that if he, Uncle Charlie, had the balls to risk his neck on such a mission, it would bolster the men's confidence in FMS.

"Sorry that the answer must be negative," Nimitz replied.[5]

In San Diego, the USS *Flying Fish* (SS-229) and the USS *Redfin* (SS-272) were being prepped to conduct the comparison tests of the other sonar devices under consideration by the Navy. In San Francisco in late February, the *Bonefish*, after shakedown and testing at sea, participated in an antisubmarine exercise with Navy pilots off Monterey, California, after which she returned to Hunters Point. On her arrival, Lawrence received his departure orders.

Sarah and Boo had train tickets for their return to Atlanta, where they would wait for Lawrence's final homecoming from the war. He bade his family a tender, tearful, and agonizing good-bye. His concern over Sarah's pregnancy during the long train trip home was eased somewhat by the fact that she'd be accompanied by several Navy wives, some of them pregnant, also returning home. To sustain him during the long, arduous days of patrolling in the months to come, Lawrence would draw on his memories of the days and nights he had spent with Sarah and Boo in Atlanta and San Francisco. He was still excited about having "Junior," due in August, waiting for him when he returned home. Sad as the separation was, a letter from his parents lifted his spirits. He quickly wrote back, saying:

> [I appreciated your letter] considering that we are all a <u>bit</u> (to
> say it mildly) down in the dumps still from having to leave our
> families so recently. It was really hard to leave them you may be

*sure. I think I can say that for all the boys, but I know I can say
it with ten times over for myself.*

*[Boo] will probably disappoint you about [not] talking of
the zoo, oceans, ships, etc. I think she is much more impressed
with her "bow dog" at home, to hear her tell it. She will also be
overjoyed, I think, to get back to her tricycle. . . . Not that I still
don't think she's the world's cutest little girl! If "Jr." turns out to
be half as much pleasure and joy to us as Sarah has already
been, he'll be far more than just welcomed, that I know.*

*. . . I've been told that present plans are for this to be my last
patrol, and that I'll probably go to Admiral Lockwood's staff
upon its completion. Hope we can make this patrol a good one.*

*Also, my Navy Cross is here, now, merely waiting for a day
when the admiral can find time to have the appropriate
presentation ceremony. I'm awfully glad that it has come before
we depart on patrol. Now I can send it home to Sarah before we
leave, and you can see it too—even if it is about the least gaudy
and fancy of all decorations. Sarah will also no doubt show you
the Bronze Star received for the patrol before this last one. . . .* [6]

The *Bonefish* sailed for Pearl Harbor at the end of February. During
the voyage Lawrence began writing a long letter to Sarah.

Most precious wife,

*It is approaching a week since I last saw you and Boo—that is
what the calendar says. If it weren't for the calendar, though,
I'm sure I'd have no clear idea of how long it has been except
that it has been at least months and for far too long. Again we
are facing the mere beginning of a long period of operation, one
which seems hopelessly long from where we are now—so long
that its end is only something to conjure . . . theoretically. I
know that my only hope is to remember that once before we
were faced with the same situation, and eventually, some way,
time passed. . . . [A]t the moment the time is dragging, oh so*

slowly, and the pain of not being able to return to you . . . is like an open wound in my very heart. . . . I don't believe that I have ever been so terribly impatient for this war to end as I am now. And that means only to be able to return with honor to live a normal life with you and Boo—and Junior/Virginia—to do only those things we can do together, to try to show you I love you as well as tell you. . . .

Coming down to earth a bit . . . For our trip, so far, I've been sleeping and reading for the most part. Reading some radio [technical manuals], a little math, a short history of China, a mystery and some short stories about, of all things, fishing. The first day, I got sea sick again and haven't felt really good or had a decent appetite. . . . All in all, I guess no one has been too enthusiastic about leaving. Maybe it will be good for us to get back out where we will have to think of something besides ourselves!

Gee, Angel, it will be wonderful if by some . . . streak of luck I really can be there on Jr.'s/Va.'s birthday. I do hope so with all my heart. Who'll take his picture if I'm not there to do it?

By for now, dearest sweetheart. Once again, here is all my deepest love to you and Boo. . . .

Lawrence

———

The *Bonefish* arrived in Pearl Harbor on March 2. Edge and his crew immediately began an intense ten-day period of training for her next war patrol. Her FMS received a thorough going-over by technicians, who pronounced it ready for unlimited operation. Back at sea and on the way to Guam, Edge conducted drills and training exercises to maintain peak performance. Blue skies and towering cumulus clouds kept the *Bonefish* company as she made her passage westward, the distance noted only by the change of time zones on the ship's chronometers and her line of advance marked on navigation charts. Though the Pacific west of Hawaii was for all intents and purposes an American lake, the lookouts and watch standers kept their eyes open for the periscopes of Japanese submarines. The enemy's big I-class boats still roamed the Pacific, though in small numbers. Most had been pressed into service as transports to deliver supplies to what remained of those troop garrisons on islands bypassed by U.S. forces and left to die on the vine. Most of the emperor's smaller, short-range RO-class subs had been sunk, many by U.S. subs.

As the *Bonefish* steamed westward Edge could only speculate on what was in store for him and his crew at Guam. Scuttlebutt said it would likely be the same thing that had awaited the other FMS-equipped subs that had preceded the *Bonefish* to Pearl Harbor. As always, scuttlebutt provided the answers that SubPac's senior officers refused to divulge ahead of time. It was said that in Guam, the *Bonefish* would receive orders to map enemy minefields for the Third and Fifth fleets. But there was something else, too. Something about an invasion of Japan by submarines. Was that true, or just a figment of someone's overactive imagination? Someone who had caught a glimpse of Mush Morton's ghost?

PART TWO

The Hellcats

PART TWO

CHAPTER NINE
An Operation Called "Barney"

A s the planning for the raid gained momentum, Lockwood still didn't know for certain whether FMS sonar—and not some other gadget—would ultimately be the tool submariners would use to foil the mines guarding the Sea of Japan.

He complained to his correspondents at the Navy Department that he had been trying to move heaven and earth to get a reliable mine detector into submarines. The uneven performance of the *Bowfin*'s and *Tunny*'s FMS had made him uncertain about FMS's future, if not angry that after struggling to get hold of a reliable mine detector he still didn't have one that was foolproof. "Lack of such a gadget leaves only the old method of under-running the fields, not a particularly inviting solution when mines can be set at most any depth," he wrote.[1]

Lockwood also complained about Admiral King's denial of his request for a relaxation of the restrictions that had been placed on publicity regarding submarine operations. Lockwood had always insisted that his submarine force operate in the shadows and that any information released to the public about its ships, missions, and personnel would be too risky and might help the Japanese to bolster their antisubmarine defenses. The term "silent service" was an apt description of the force. But now with the end of the war just over the horizon, Lockwood had had a change of heart. He wanted to release information about the sub force that would bolster its image in the public's mind and educate people about the genuinely astonishing accomplishments and sacrifices of the force. It would also inform the public about how dangerous submarine operations were and set the stage for the honors that would be owed the men who had participated in the raid. Lockwood said that publicity would blunt some of the "ballyhoo from our allied services. The remark [made by a general officer] about the AAF [Army Air Force] making possible the invasion of Europe still rankles." A letter from King disapproving any and all submarine publicity had settled the matter, at least temporarily, Lockwood grumbled. When his raiders returned from the Sea of Japan things would look quite different.

In his relentless drive for advanced prosubmarine gear, Lockwood pushed hard to speed up delivery of those cutie homing torpedoes. Since his last push, he'd learned that the Royal Navy had captured a GNAT (German Naval Acoustic Torpedo), the homing torpedo that had been used so successfully against Allied shipping by U-boats. An autopsy of its innards had disclosed important facts about its operation, especially its exploder. Lockwood urged BuOrd to appropriate the exploder's design for use on U.S. Navy torpedoes.

It was mid-March and Lockwood had four FMS-equipped subs available for his mission: *Spadefish*, *Tinosa*, *Tunny*, and *Bowfin*. A fifth sub, the *Bonefish*, was on her way to Guam, and a sixth sub, the *Flying Fish*, in San Diego, would be available after her duties as a test bed for the competing sonar systems. (The other test bed, the *Redfin*, would be released for regular war patrols after performing similar duties.)

Lockwood had a decision to make and he had to make it now. Lacking anything better, he'd planned the mission around FM sonar. Working sets had been installed in subs. There was no guarantee that the other mine detectors due for competitive testing would perform any better than FMS. Even if they did it would take weeks, maybe months to rip out the FMS installations, replace them with another system, and conduct tests under combat conditions. It was too late for that. Those were weeks and months Lockwood couldn't afford to waste. Given the U.S. advance across the Pacific, the invasion of Iwo Jima that had begun in February, and the invasion of Okinawa scheduled for April, the noose around Japan was tightening by the day. If the mission had to be delayed until July or August it might as well be scrubbed right now. His instincts told him that he'd made the right decision, that none of those competing sonar systems would best FMS. He'd keep an open mind if it would help garner support for the mission from the skeptics back in the States who, for various reasons, were pushing their own pet versions of submarine mine detectors. But he was fully committed to FMS and not about to switch.

With this decision, Lockwood and Voge plunged into the details of the mission. They debated how many subs to employ for the raid. Two other subs were undergoing overhauls and FMS installation on the West Coast. They were the USS *Crevalle* (SS-291) and the USS *Seahorse* (SS-304). It seemed reasonable to send in as many subs as possible, perhaps as many as a dozen, depending, of course, on how many FMS sets were available for installation. This action would force the Japanese to spread their antisubmarine forces over a large area to hunt for many subs, instead of concentrating their efforts on just a few in a small area. The final decision about numbers would have to wait until after the sonar competition and until it was determined how many sonar-equipped subs were available.

Lockwood had already decided that the raiding subs, helped along by the swift-inflowing Kuroshio Current, would enter the Sea of Japan submerged via the Tsushima Strait south of Kyushu. His decision had been made based on all the information collected about patrols, tides, currents, and minefields. Lockwood and Voge had even identified a starting line for the operation, located south of the entrance to the Tsushima Strait. It was time now for an FMS sub to map that area so the raiders would know where the mines were and how they were lined up in the mouth of the strait.

Lockwood and Voge chose La Pérouse Strait in the north for an escape route. With its outflowing Kuroshio Current and mine-free channel used by the Russians, it would be ideal for the purpose, provided the USSR didn't declare war on Japan, which would cause the Japanese to seal the entire strait with surface mines. Lockwood worried that if the Japanese sealed La Pérouse while the raid was unfolding, the subs would be trapped inside the Sea of Japan until they ran out of food and fuel and were hunted down and sunk. That possibility seemed remote, as the Russians were in no hurry to open a second front in the Far East. In that case, Lockwood was confident that the subs, using stealth and speed, could make their exit dash out of La Pérouse unmolested.

Lockwood and Voge knew that FM sonar was far from perfect, perhaps no more than eighty percent effective. They knew, too, that some of the minefields the subs might have to breach were part of the old antisubmarine fields laid by the Japanese in early 1941 and since scrambled up by storms and time. Regardless, there was little they could do but make the best of what they had; no amount of hand-wringing over these issues would ensure total success. Lockwood the doer was committed to his plan and to seeing it through no matter what. A keen observer of operations in the European theater as well as the Pacific theater, he had long ago adopted a credo espoused by General George S. Patton: "A good plan, violently executed now, is better than a perfect plan next week."

Despite Lockwood's determination to launch the mission before summer, he still had to solve the lingering problems plaguing FMS. He wasn't happy with the diminished range sensitivity exhibited by some units, especially under less than favorable sea conditions. Lockwood believed that the problem of diminished sensitivity could be solved if the subs were to underrun the minefields at a depth—say, greater than 180 feet—where they'd encounter quiet water conditions that would enhance FMS's sensitivity. His reasoning sounded good, but in practice was hard to prove in the turbulent environments in which submerged submarines often had to operate.

The arrival at Saipan of the *Tinosa* and *Spadefish* proved how much work still needed doing. The *Tinosa*'s FMS did not perform satisfactorily even with help from Malcolm Henderson with his circuit testers and amp meters. The results left her CO, Richard Latham, baffled. He'd turned in a satisfactory mine recon of Okinawa with the very same equipment. While Henderson fiddled with it, Lockwood turned his attention to the *Spadefish*. She and her new skipper, Commander William J. Germershausen of Baltimore, Maryland, were scheduled to depart on a war patrol, and Lockwood wanted to be certain that the *Spadefish*'s FMS was in top form.

After long hours spent in cramped, hot quarters tuning the *Spadefish*'s FMS, which had been updated with the very latest electronics from UCDWR's labs, Henderson declared it ready for unlimited service. Germershausen, itching to go, received a set of sealed orders from Lockwood that included an important set of instructions to reconnoiter the southern limits of the Tsushima Strait's minefields. Germershausen also had orders to pick up any Japanese survivors from torpedoed ships found floating in the ocean, especially around Tsushima, who might have information of value to the impending mission.

With that settled Lockwood returned to the *Tinosa*'s sonar problems. Discouraged at having to wrestle with a balky FMS installation that had once worked so well, Lockwood worried that all the effort he, his staff, the scientists at UCDWR, and the skippers and sub crews had expended to turn FMS from a bright idea on paper into a working reality could yet go down the drain. His natural optimism had deserted him. And so had his luck. No one would blame him for taking a scuttling sledge to the damn thing. He might have even been on the verge of wielding one aboard the *Tinosa* when all of a sudden his luck changed.

In quick succession the *Tinosa*'s retuned FMS turned in an excellent performance at sea, the best of any sub so far; and the *Tunny*'s skipper, George Pierce, radioed that they'd had extraordinary success charting mines in the East China Sea. His message read, "Top Secret. To ComSubPac. Completed passage through the East China Sea mine field. Have charted rows of mines spaced about one thousand yards apart. Position about 170 miles north by west of Okinawa. Am now in the East China Sea and have charted 222 mines in the vicinity of Lat. 29-20 north, Long. 127-10 east [the Ryukyu Island chain south of Kyushu]." At the end of the message Pierce added what he knew Lockwood desperately wanted to hear. "FM Sonar gear running like [a] sewing machine."[2]

Pierce confirmed what Lockwood had surmised all along: Minefields had indeed been sown south of Kyushu, and they were responsible for the loss of at least four, and perhaps as many as six, U.S. submarines. Pierce had proved that a submarine equipped with FM sonar could penetrate these deadly fields, map them, and count them like so many rows of pumpkins. Lockwood was ecstatic; his luck hadn't deserted him after all. He couldn't wait to trumpet Pierce's successful recon; it would be a confidence builder for all those who still had doubts about FMS. Yes, it would still require a lot of arm-twisting and cajoling, but Lockwood was good at that.

In earlier correspondence with those officers in charge of FMS training on the West Coast, he had laid on the line what had to be done to convince the naysayers. ". . . [C]onfidence must be built up by familiarity with use and knowledge of enemy mining systems and limitations. Dr. Henderson and I have been doing it here [at Guam] and while I like to get out in the boats, eventually there will be more than I can handle. Where we find COs opposed to use of FM Sonar, we will have to relieve them [of command], for no one will do a proper job of mine detecting unless he has confidence in his gear and has plenty of guts. . . . Each sub should have at least 3 good operators. . . . We've got to get that Tsushima job done—and soon."[3] Finally, he had a solid success that would prove how good FMS was at mine detection.

Lockwood's report to Admiral Nimitz on the *Tunny*'s success elicited CinCPac's congratulations and a personal message to Pierce and his crew for a job well-done. Personal recognition by Nimitz was always a terrific morale booster and, in this case, one that polished Lockwood's brass in the bargain. Basking in the glow of Nimitz's praise, Lockwood tried to wheedle himself a ride aboard the *Spadefish*, about to depart on her mission to Tsushima, claiming that it was imperative that he go along to size up her performance. He hurriedly sent his request to Nimitz, only to find that the admiral had departed for Washington. With no time to waste he pleaded his case to Nimitz's chief of staff and war plans officer, Rear Admiral Charles H. "Soc" McMorris. Speaking for Nimitz, McMorris turned him down flat. "You know too damned much about our future plans," he said, in a rare instance of a rear admiral issuing orders to a vice admiral. "But, Soc, if we come to grief on this mission—which I'm sure we won't—there'll be no prisoners." McMorris shook his head no and that was it.

McMorris's refusal wasn't unexpected. Lockwood, the armchair submariner, accepted that with the war's end looming he'd probably never get to make a war patrol nor fire a torpedo in anger. What rankled him, though, was that his friend James Fife had wrangled a patrol aboard a submarine operating off Subic Bay. How he had done it was a mystery to Lockwood and the cause of deep envy. Here was ComSubPac, who couldn't even order himself aboard one of his own subs, while Fife was having the time of his life off the Philippines. Riding a sub to test FMS was hardly a substitute for the real thing.

In the meantime two problems that had been pushed aside for more pressing matters vied for Lockwood's attention: the upcoming test in San Diego of rival sonar detectors scheduled for late April; and finding an experienced submariner, an ex-skipper, to take over the training, planning, and execution of the Japan Sea mission. Selection of an officer for this assignment would relieve Lockwood and Voge of the enormous workload threatening to grind them down under the already formidable task of running the submarine war.

Lockwood directed his chief of staff, Commodore Merrill Comstock, to find an officer who could take up the load. Comstock tapped ComSubPac assistant operations officer Commander William Bernard "Barney" Sieglaff for the job. Sieglaff was the former skipper of the USS *Tautog* (SS-199) and USS *Tench* (SS-417) and had thirteen confirmed sinkings to his credit.* Sieglaff was one of those quiet, resourceful, and self-motivating officers who often work behind the scenes of big and important undertakings that are successful because of men like him. With his appointment in late March the Tsushima Strait breakthrough finally had a name: Operation Barney.

Even as Operation Barney's load shifted from Lockwood's and Voge's shoulders to Sieglaff's, the *Seahorse*, skippered by Commander Harry H. Greer, Jr., and the *Crevalle*, under Commander Everett H. Steinmetz of Brooklyn, New York, arrived at Guam. Lockwood put the two subs through four days of intense FMS training under his and Sieglaff's supervision. As expected, the gear in both subs needed a tune-up to meet

---

*At the end of the war the *Tautog* would claim the record for the most ships sunk, with twenty-six.

Lockwood's and now Sieglaff's exacting standards. Lockwood wasn't discouraged anymore. He knew they were closing in at last on solving the few remaining problems with FMS. He was also encouraged by a genuine improvement in the attitudes of FMS sub crews and also in their mine-hunting skills, which in turn bolstered everyone's confidence.

In late March Lockwood dispatched Greer in the *Seahorse*, as he had Germershausen in the *Spadefish*, to sniff around the southern limits of the Tsushima mine barrier. He did this to double up on the mapping and to ensure its accuracy. Next he sent Steinmetz in the *Crevalle* off to patrol along the China coast and to probe the southern Tsushima boundary.

The *Bonefish* nosed into Apra Harbor, Guam, on April Fool's Day. Lockwood and Sieglaff wasted no time climbing aboard for FMS tests off Orote Point. Lockwood, enthused as ever, cast a keen eye on Edge and his ship, which arrived in spotless condition. He remarked that Edge, an electronics expert, was keen to wring out his ship's sonar equipment and that he was self-assured and all business. Lockwood knew through scuttlebutt that sometime in July Edge would receive orders to the electronics desk at the Navy's Bureau of Ships (BuShips), not ComSubPac, as Edge had assumed. Before this happened, however, Lockwood wanted him for Operation Barney. He was eager to watch Edge, one of the only sub skippers in the force who had been trained in the arcanum of circuit boards and resistors, put FMS through its paces.

Even before the *Bonefish* could get under way at dawn, problems arose. A cable had pulled out of her FM sonar head due to faulty switches that controlled the rotational limits of the head shafting. Lockwood kept his anger in check while he chain-smoked, waiting for the technicians to complete repairs, not pleased with the delay nor with the fact that the alleged experts who had gone over the ship's equipment upon her arrival hadn't discovered the problem. After repairs the *Bonefish*'s performance that day bested any that Lockwood had seen. Coupled with Edge's knowledgeable touch on the controls, it tempered the admiral's earlier upset. Despite FMS's temperamental nature, when it was handled properly, as it was by Edge, it worked beautifully. Lockwood was also pleased to see that Edge didn't display the skepticism and cynicism about FMS that had affected so many of his peers. Lockwood could cheerlead all he wanted for his plan, could talk it up day and night, do a sales job on it that would hopefully convince even the most skeptical among them. In the end he had to prove that it would work. And he had to hope that events beyond his control didn't scuttle his efforts. More than anything else, he feared that the loss of one of the three FMS subs now on patrol in mine-infested waters south of Kyushu, or those that would follow, would sink Operation Barney before it ever had a chance to get started. So far early reports from the skippers of the *Spadefish*, *Crevalle*, and *Seahorse* indicated that, like George Pierce in the *Tunny*, they were having success locating mines. Watching Edge demonstrate his expertise of the gadget and seeing his confidence in its abilities was all the evidence Lockwood needed to maintain his belief in FMS and Operation Barney.

When Edge returned from his outing with Lockwood, he started a letter to Sarah.

> *We've been operating pretty steadily, starting early and ending late every day. Admiral L. has ridden with us one day, and he, in the several times I've seen him (lunch once, dinner once, here on the boat twice, etc.), he has turned out to be much nicer and more interesting than I had previously thought or imagined him to be. But I'm still not sure that I'll be on his staff after all. . . . I'm not worrying about it one way or the other, they'll probably change their minds this way and that several times before I get back from patrol.*

A day later, after completing another workout, he wrote:

> *Well, every day something new. This [next] patrol may not be my last after all! So the admiral happened to indicate today. He in fact doesn't seem to have heard of all the things the other folks of his staff have been telling or are planning. So now I*

don't know what to think. Even so it appears that this one and
another [patrol] will be all. . . .

Being infected with Lockwood's enthusiasm for FMS may have been what Edge needed to convince himself that the mission he would be undertaking would help bring the war to an end and speed his return to Sarah and Boo.

CHAPTER TEN
The Minehunters

The *Bonefish* departed Guam on April 6, after Lawrence Edge received sealed orders for a seventh war patrol in Area Nine, the Goto Retto and Quelpart Island region southwest of Kyushu. His orders included instructions to operate with the *Seahorse*—if feasible, given Greer's tricky assignment—and the *Crevalle*. Edge also had orders to locate a minefield in an area southwest of the Danjo Gunto, a small group of islands west of Kyushu. The orders didn't specify a date for carrying out the mine recon, instead leaving it to Edge's discretion.

In private moments during the voyage Edge's thoughts turned to Sarah, Boo, and "Junior." Time permitting, he wrote letters to Sarah in which he revealed deep, personal feelings about the war he so much wanted to survive.

Though Lockwood had delighted in Edge's optimism regarding FMS, and his impressive bearing and self-assured manner, Edge's true feelings, which he put into his letters, had been tempered by the great distance and duration of separation from his wife and daughter, and by the deaths of men with whom he had served and had shared, as men in combat do, his deepest fears and longings. Those fears and longings, and a sadness bordering on melancholy, are apparent in photographs taken of Edge during the period leading up to Operation Barney. He appears drawn and tired, world-weary and anxious to have the war over.

Most Darling Love,

. . . I write every day . . . telling you how much I miss you, and Boo and "Jr.", and how dearly I love you and them. . . . It is

*particularly true at the beginning of the patrol . . . because . . .
anxiety is probably at its greatest, in spite of trying to imagine
that we're awfully brave, well rested, and ready to go and make
another killing and Navy Cross.*

*To me this has been the hardest patrol of all . . . and I
suspect it's mostly because I have so recently . . . been with you
that I miss you more completely . . . than ever before. Now that
I know full well what war is like, at least in submarines, and
know that I don't really like it and never will (war, I mean) . . .
so I'm mainly conscious of . . . having to be here instead of there
with you. Feeling the war is approaching the downhill side, as
far as time is concerned, is no help to any of us either. Rather
there's the feeling that we've been lucky enough to survive so far;
it would be such a shame not to last for the remainder and thus
live through the whole thing.*

*. . . [S]ome of the heaviest fighting (if not the worst of it) of
this theater is yet to be done and probably even heavier loss of
life is yet to come than has already. That being the case, why
should we in submarines sit back yet and say, in effect, that we
are through and our part of it is over? That applies especially to
me and to most of us now on the Bonefish, since few of us have
been fighting long enough to say that we have about done our
share. . . . So, by rights I should be going on patrol ready to take
all reasonable risks and chances the more of them [there are]
the more the difficulties which confront us.*

*. . . I want so strongly to return safely and bring the boat
and the whole crew back safely that one of my big fears is that
I'll let that desire interfere with what is my real duty to the
winning of the war and to those who have already given so
much more than I in either risks or sacrifice. That . . . is one of
the worst phases of war to anyone actually engaged in the
fighting: the mental conflict between what he really <u>wants</u> to do
and what he believes he <u>should</u> do. Maybe it just means I'm
more scared than ever before, I don't know. Probably I
shouldn't even be writing these things at all, because somehow I*

suspect it would be better not to admit them even to myself. . . .
[S]tatistics still say we have a pretty good chance of getting back
all right. I just hope I don't disgrace myself in the eyes of other
skippers, the boss [Lockwood] or my crew. . . .

I miss [you] with the tremendous longing that all but
overpowers me at times. . . . I can hardly wait to have little
brother join us and to join you three myself. . . . I just want to
be with you all so much that I don't want to be out here even a
little bit, especially if there is the least doubt that I'll not return
to you. . . .

Lawrence[1]

———

Plowing through heavy green seas at two-engine speed, white spume exploding over her bull nose, the *Bonefish* cut through the Ryukyus for Area Nine, entering the East China Sea just after dawn on April eleventh.

Edge, alert and decisive, his submariner's sixth sense on guard for any indication that a potential problem was brewing, kept a firm hand on both his ship's and his crew's pulse as they approached Japanese-controlled waters. If he felt a quickening of those pulses it was because, operating so close to Japan, anything could happen. He made certain that the crew maintained a hair-trigger posture, ready to react instantly to an emergency or sudden contact with the enemy. The *Bonefish* rarely needed her crew's assistance to operate with complete reliability and was in fact operating as if she had no need for them at all. That wasn't completely true; every man aboard played an important role maintaining the submarine at the highest level of combat efficiency.

At night, on the surface in the East China Sea, the officer of the deck, the quartermaster of the watch, and four lookouts stationed on the bridge peered out through binoculars at the dark stretch of water, searching for enemy ships and planes. As the *Bonefish* swept toward her patrol station the radarmen studied the bright green pips blossoming on the radarscope like stars in a galaxy each time the ship's powerful SJ radar swept around. The pips indicated scores of night-fishing sampans and sailboats, while ghostly greenish background clutter called "grass" indicated small, grubby volcanic islands or fast-moving rain squalls. The radarmen, searching for targets worthy of a torpedo or two, evaluated each pip, dismissing some, keeping an eye on others.

After sampling the array of contacts on the radarscope in the conning tower, Edge backed down the ladder into the control room, where the chief of the watch acknowledged that everything was under control. Edge conferred with his executive officer over a chart of the patrol area. The exec confirmed that the *Bonefish*'s present course and speed conformed to Edge's orders. He also pointed out that the steering effects produced by the strong local currents required constant course corrections. Detailed information on those currents and other essentials pertaining to navigation in waters around southern Japan had been published in the ship's dog-eared copy of the *Coast Pilot*. Area Nine was strewn with dozens of small islets and rough-bottomed coastal shallows similar to those in waters the *Bonefish* had patrolled off the Philippines and that posed a potential hazard.

Moving aft, Edge passed the radio room, banked with transmitters and receivers, and entered the crew's mess, where off-duty sailors gathered to eat meals, drink coffee by the gallon, play cribbage and backgammon, whine, complain, and talk about what they were going to do when they got home after the war. Other favorite topics for discussion were the Navy, and, of course, women.

Continuing aft through the crew's berthing compartment, past snoring off-duty sailors sprawled in their bunks, Edge entered the booming engine rooms. He acknowledged the motor macs on watch and received a circled-thumb-and-index-finger status report from the engineering officer. The men used sign language to communicate; talking over the deafening roar of diesel engines running at flank speed wasn't possible.

Farther aft, in the maneuvering room, Edge chatted easily with the electricians' mates controlling and monitoring the ship's electrical load at the propulsion control console. He surveyed the console's solid wall of amp meters and gauges and checked the lineup on the main propulsion control levers. Nodding approval, he moved on to the after torpedo room. He found, as he had during an earlier tour of the forward torpedo room, a gang of cocky and profane torpedomen itching for action. Sharing their living and bunking spaces with a load of hulking three-thousand-pound, twenty-foot-long torpedoes, the sailors had developed a genuine affection for them, patting their blunt noses as if stroking a faithful dog on the head. Torpedomen assigned to the after room saw less action than their mates up forward. Regarded as tail-end Charlies, they never failed to ask Edge, whom they all admired and respected, if, when he had a target, he could try to fire the stern tubes first. The easygoing Edge would try his best.

After sunrise, as the *Bonefish* drove northwestward toward the Goto Retto, a fast-approaching plane lit up SD radar. It would be on top of them in minutes.

Edge didn't hesitate: "Clear the bridge!" The officer of the deck hit the diving alarm, honking it twice.

Seawater surging into her ballast tanks, the *Bonefish* nosed down, rigged-out bow planes on full dive. Water streaming past the plunging submarine sluiced into the superstructure's voids, chuckling up the sides of the conning tower and periscope shears and over the sealed bridge hatch.

"One-five-zero feet. Eight-degree down bubble," said the diving officer, confirming Edge's orders. The *Bonefish* dived fast, clawing for depth and the protection it offered.

Edge called for a range estimate on the airplane before the SD mast went under. "Five miles, Captain, maybe less." He altered course away from the telltale scar of foam and bubbles the *Bonefish*'s dive had etched on the surface.

A water hammer rattled pipes and valves in the forward part of the ship. The hydraulic system moaned as the planesmen finessed the bow and stern planes just so to level the sub out exactly at Edge's ordered depth. "One-five-zero feet, Captain." Edge acknowledged the report and nodded approval.

Though the airplane had likely come and gone without dropping bombs or depth charges, Edge ran submerged all morning, until it was time for a cautious look-see.

"Sonar?"

"Nothing in the vicinity, Captain."

"Up scope." On haunches, Edge waited for the periscope to appear, snapped down the folded training handles, and rose with it. Arm draped over one of the handles he muscled the scope around 360 degrees, searching sky and sea for intruders or, better yet, targets. It took only seconds to make a full circuit. "All clear. Down scope." It was time to make tracks for the patrol station. "Control, prepare to surface."

Edge, the officer of the deck, the quartermaster, and four lookouts, with binoculars slung around their necks, waited at the foot of the ladder to the bridge.

"Surface the boat," Edge ordered.

Three hoots of the Klaxon resounded through the ship, followed by the roar of high-pressure air dewatering the main ballast tanks. The *Bonefish* shuddered and nosed up. As the depth gauges unwound, the diving officer called out the readings until the boat had shouldered out from under the sea. Moments later four diesels, rolling on air starters, erupted, spewing smoke and cooling water from their exhaust ports. With two engines charging batteries to ensure that the ship had a "full can" for the next submergence, the staccato throb of the two on propulsion rose several octaves. The screws took a bite and the *Bonefish* swung onto a course into the East China Sea.

After sunset on the twelfth, the *Bonefish*, patrolling south of Quelpart Island, exchanged recognition signals with her pack mate, the *Seahorse*, via SJ radar. Radar pulses keyed via Morse code allowed submarines to communicate on a secure channel. After an exchange of information, which included Edge's report that they'd encountered and destroyed several floating Type 93 mines with gunfire, Greer gave Edge permission to patrol independently, as the area they were in, laced with islands and shallows patrolled by the Japanese, would impose severe limitations on a coordinated attack. As well, the *Seahorse* would be unavailable for a few days while she conducted that mine recon for Lockwood around the mouth of the Tsushima Strait. When the *Bonefish* and *Seahorse* parted company Greer and his men had no idea what the Japanese had in store for them.

The next day the *Crevalle* showed up to exchange information with the *Bonefish*. In between torpedo attacks, some successful, some not, Steinmetz had been busy reconnoitering minefields. Steinmetz thought it was dangerous having three subs that were operating in one corner of the patrol area rendezvous. He disliked what he called "this dog-sniff-dog stuff" and the "mad flail," as he put it, to make contact. The two subs parted company and moved on.

Shortly, a grim-faced radioman handed Edge a decoded message from Fleet Radio in Pearl Harbor: President Roosevelt had died in Warm Springs, Georgia.* Details were sketchy, as was information about the new president, Harry S Truman, whom some of the sailors had never heard of. If any of the men had wanted to take a moment to reflect on FDR and what effect his death might have on the outcome of the war, radar contact with a small echo-ranging patrol boat scrambled them to action. Though it was a very dark night and the patrol boat was only a hazy blur, Edge believed that he had a target worth a torpedo.

"Stand by tubes One, Two, and Three. Set depth four feet. Gyro angles zero-one-one, zero-one-two, zero-one-three." The spread of gyro settings guaranteed a hit as the patrol boat advanced left to right across the *Bonefish*'s bow. "Fire One!"

The firing key operator smacked the firing plunger. "Number One fired electrically!"

A torpedo roared from its tube, turbine lighting off like a buzz saw. Two more followed at ten-second intervals from the forward tubes—not the stern tubes the tail-end Charlies had hoped for.

Edge pulled away at full power. Aft, in the sub's boiling wake, the tin fish appeared to run hot, straight, and normal when suddenly all three broached like playful porpoises. Instead of warheads exploding against the enemy ship's hull, there was only a crushing silence. Alerted by the broachers, the patrol boat turned on her heel and charged back down

---

*The president died in Georgia on April 12. For the *Bonefish* crew patrolling west of the International Dateline, he died on April 13.

their damning wakes for the *Bonefish*. Edge, hauling out, led a merry chase until the Japanese skipper broke off. Edge decided against a follow-up attack; he couldn't justify shooting more torpedoes at a target itching for trouble. Instead, he shaped a course to patrol an area across the southern traffic lanes south of the Tsushima Strait, where he hoped to find loaded ships bound for Japan.

Two days passed uneventfully. The routine of a so far unproductive patrol was broken only by a parade of small sea trucks and sampans and the sound, always like distant thunder, of heavy depth charging. Perhaps it was the *Spadefish* or the *Crevalle* getting a working over by the Japanese. It wasn't hard to picture eighty-some sweat-drenched men breathing stale air trapped inside a submarine's hull, counting the seconds until the next drop, praying that it wouldn't be on top of them. The thunder slowly faded like a departing storm, the outcome unknown until the next rendezvous—or unanswered call—between subs. The *Crevalle* showed up around midnight on the fourteenth, proving that she'd not fallen victim to the Japanese. There was no sign of the *Spadefish*.

Edge took advantage of a lull in the action to start a letter to Sarah. Writing in short spurts, he had time to complete only a small portion each day and used dashed lines (- - - -) to indicate a time break between each part.

> *Lovely, dearest wife,*
>
> *What can I say but that I love you with all my heart and soul? . . . That I miss you with a constantly aching heart? . . .*
> *The patrol is moving along but slowly. Events take place but practically none of them are encouraging. Actually there have been few dull days and too few dull moments, though I think we'll all be glad when this patrol is over. Luckily (and I haven't mentioned it to you before) it is to be a short one, because of a little special mission assigned. We are thankful for that, anyhow. I know you will jump to the conclusion that we've had*

*a highly successful patrol when you receive my cablegram upon
our arrival at base, but unless things change radically from
their ways so far, it will be a false conclusion. That is one reason
we'd just as soon start back now: this area is unpleasant and
still there seems little opportunity even to get combat pins and
stars for the boys who don't have them.*

— — — — — — — — — — —

*Angel mine, I still think and dream of you too much of every
hour of every day! How the time drags! I wonder what you and
Boo . . . are doing. Already I can't help thinking how much Boo
will be changed next time I see her. . . .*

*In a way I'm surprised to calculate that it has been only
roughly two months since I last saw you . . . that makes Jr. only
about (-)3 months old now! . . . Tell him again for me, though,
that he had better take the very best care of you. . . .* [2]

On April 15, the *Bonefish* received a lifeguard assignment in waters
around the approaches to the Tsushima Strait. Hundreds of B-29s flying from bases in the Marianas had been hitting targets deep inside
Japan, while suffering heavy losses to enemy fighters and flak. If the
crew of a B-29 went down, they stood a good chance of being captured if
not killed by the crew of one of those ubiquitous Japanese sea trucks or
sampans.

Patrolling near Kyushu, the *Bonefish*'s APR radar-detection gear registered a marked increase in Japanese radar emissions. Earlier, Greer had
reported to Edge and Steinmetz that a similar increase had been register-
ing on the *Seahorse*'s APR. The Japanese navy, despite its depleted state,
had recently installed on its patrol boats a crude radar system whose
emissions mocked submarine SJ radar. More than one sub skipper, de-
coyed into thinking he had APR contact with a U.S. sub, blundered into a
waiting patrol boat. The buildup of radar emissions was a clear indication
of increased antisubmarine patrols in the southern Kyushu region. Steinmetz's earlier quip that the dog-sniffing-dog routine was a prescription for
trouble wasn't lost on Edge as he moved toward the Tsushima Strait to
assume lifeguard duty.

———

The **Bonefish** patrolled a thirty-mile triangle in the southern approaches to the Tsushima Strait,* radiomen alert for any messages reporting downed B-29s. It was hard to maintain a steady guard on the lifeguard frequency, as the *Bonefish* had to bob up and down like a cork to avoid pesky patrolling Japanese planes. Toward noon, during a long, uninterrupted stretch of surface patrolling, a lookout sighted a heavy puff of black smoke on the horizon, and near it, a large flying boat that looked like a U.S. Navy PBY. The plane disappeared into the haze but the smudge of smoke lingered. Smoke on the horizon usually meant that a ship, or better yet, a convoy, might be just over the hill. Edge, eager to find out, ordered, "All ahead flank!"

The motor macs kicked the throttles wide-open, spooling up the diesels. The *Bonefish* heeled around in the direction of the smoke. Despite distant aircraft contacts on SD radar, Edge was determined to stay on the surface as long as possible to find the ship before it disappeared. With luck he could dash in, make contact, and, if he did, submerge and fire torpedoes. Yet after a half hour of hard running toward a seemingly endless horizon, there weren't any stick masts or smoke, much less ships, in sight. Had the lookout been fooled by a mirage? It happened all the time. Edge didn't blame the lookout; the sea and sun often played optical tricks on seafarers. Still, the young sailor was adamant that smoke was what he saw, not a mirage. But a high periscope search over the horizon revealed only sampans and seagulls, not ships. After an hour spent searching for a phantom in the area of the sighting, Edge was about to turn back when another lookout spotted a large oil slick a half mile off the port bow, and something floating in the water.

Hauled aboard two Jap aviators from the middle of oil slick after watching another one get out of his life jacket and submerge, not to reappear. Questioning these prisoners disclosed

———

*Here Edge may have been risking an encounter with mines; if he was using the ship's FMS to detect any, it isn't mentioned in his patrol report.

that the smoke had been from their JAKE type plane* as it crashed after being shot down by a U.S.N. PBY: the large plane seen from the bridge must have been this PBY. Our lifeguarding turned out differently from what might have been expected, but, anyhow, if the PBY had gone down instead, we would have been there. One prisoner's foot was badly smashed; the other was the non-com pilot of the plane, who stated the plane was on routine [antisubmarine] patrol! Headed NW back to patrol spot.[3]

With the two prisoners aboard the *Bonefish*, Edge had fulfilled another mission requirement Lockwood had inserted into the op orders: "When possible, capture Japs for interrogation."

In the late afternoon the next day the *Bonefish* closed in on a two-masted junk and began circling, 40mm guns manned. The junk's badly frightened crew, young and old alike, gathered on deck, waiting to be boarded. The *Bonefish*'s photographer's mate took motion pictures as she drew alongside. A boarding party armed with Thompson submachine guns and Browning Automatic Rifles motioned for the junk's crew to put over a fender. A man from the vessel clambered aboard the *Bonefish* to meet one of the Japanese fliers who had offered to interpret. The flier identified both the junk and crew as Korean fishermen. Edge doubted this was true, since the junk looked new and well equipped and was just as likely carrying contraband to the Japanese as it was netting fish. But the forlorn appearance of her crew influenced Edge's decision to let them go.

With free time, Edge wrote Sarah about the Japanese prisoners aboard the *Bonefish*, describing what one of them kept begging him to do.

We fished a couple of Jap aviators out of the water the other
day. They were none of our doing, because they'd been shot

---

*The Aichi E13A, code-named "Jake," was a three-man single-engine floatplane used primarily for reconnaissance.

*down by one of our own planes. We just happened along . . . in
time to rescue them. They assumed a little too quickly that we
were a Jap sub, so came on board all too easily. Since then, one
of them has been begging me to shoot him, every time he sees
me! He can write English a bit; so our conversations are . . .
[written down]. I have one . . . of a long-winded one I had with
him: he wanting to be killed, and my trying to dissuade him!
After it was all over he still said, "I [am] happy for death!" As
a missionary then, I'm a poop out, but I'll bring you the
conversations as a souvenir, anyhow. Of course we shan't kill
him, but are keeping him as a prisoner, our only fear being that
he'll try to [kill] himself, so all precautions are being taken as to
guarding him continuously. It's a nuisance to have them aboard
anyhow, since we really don't have the room at all.*[4]

Released from lifeguarding operations, the *Bonefish* probed for targets in waters close to the Korean coast around the Tsushima Strait, but
found only sailboats and sea trucks. Seas began building on the eighteenth, forcing Edge to give up the hunt for ships and instead set a course
for the vicinity of the Danjo Gunto minefields. The bad weather moving
south with the *Bonefish* also deterred Edge's first attempt to map the
minefields. It didn't take much imagination to picture how the heavy
seas battering the sub had stirred up all those mines. The recon Lockwood had ordered was dangerous enough without adding to it the risk of
colliding with a row of ugly horned balls swinging wildly at the ends
of their anchor cables.

Patrolling, waiting for calmer weather, Edge worked on his letter to
Sarah. He included information picked up from the daily fleet news
broadcasts and more about his prisoners.

*Well, the boys in Germany seem . . . about ready to end
things. . . . I'm impatient for the war to end. . . . I can't help
feeling that, far away as Europe is, the end of that war brings
my eventual return to you a little closer. And that, of course, is
about all any of us live for out here.*

*I wish [our troops] could land in Japan itself . . . [that] their
homeland could fall [next]. . . . Even our Jap prisoners know
that Japan is losing the war. . . . But one [of them] is still
disappointed that he didn't die—die in battle so as to bring
honor and happiness to his family! Now he thinks his family
probably believes him dead, and he is worried about their
shame and sorrow when they find out he is still living as a
prisoner instead of dying gloriously on the field of battle. . . . No
wonder it is difficult to make them stop fighting.*[5]

The twenty-first brought clear weather and calm seas. In preparation
for the mine recon the sonarmen began tuning up the FMS gear. Edge,
meanwhile, brought the *Bonefish* around onto a course that would inter-
sect the northeastern end of the mine line. As dawn broke he gave the
order to dive.

CHAPTER ELEVEN
Probing the Line

As Lawrence Edge prepared to carry out the mine recon off Danjo Gunto, the plans for Operation Barney's implementation that had been in flux for months finally jelled. Barney Sieglaff and Dick Voge settled on a basic operational plan. The mission would kick off from Guam on May 27 and employ nine subs in three separate task groups, numbered TG 17.21, TG 17.22, and TG 17.23. The groups would penetrate the Tsushima Strait individually on successive days through June 4, 5, and 6.

Once inside the Sea of Japan, the packs would operate independently in three areas. TG 17.21 would stalk along the northwestern coast of Honshu; TG 17.22 would patrol along the southwestern coast of Honshu and Kyushu; TG 17.23 would operate in an area along the east coast of Korea from Tsushima Island to the southern coast of Siberia. The mission would end on June 24, when the three task groups rendezvoused to make their exit run through La Pérouse Strait.

The nine FMS-equipped submarines included the *Seahorse*, *Crevalle*, *Spadefish*, *Tunny*, *Bonefish*, *Flying Fish*, *Bowfin*, *Tinosa*, and the USS *Skate* (SS-305). The *Skate*, commanded by Commander Richard B. "Ozzie" Lynch, a Naval Academy classmate from Alabama and friend of Lawrence Edge, was added to the group when she arrived in Guam. The *Flying Fish* was still in San Diego for the comparison trials between competing sonar-detection systems. The Operation Barney brain trust of Lockwood, Voge, and Sieglaff assumed that FMS would ultimately prove itself in tests against competing systems, and when it did, the Barney subs would assemble in Guam. As for the tactical aspects of Barney, fresh intelligence backed the use of La Pérouse Strait as an exit route. And to ensure that the subs didn't run afoul of Allied mines dropped in the Sea of Japan by B-29s, charts of those areas would be sent to ComSubPac for distribution to the submarines.

With everything in place, Voge submitted the plan for Operation Barney to the Pearl Harbor-based Submarine Operations Research Group (SORG) for an evaluation of the risks involved. SORG consisted of a group of civilians, all of whom were experts in the field of actuary science, among other disciplines. The group provided ComSubPac with objective evaluations derived from theoretical and statistical analysis of the tactics commonly employed by submarines, including torpedo attacks and methods of evading enemy antisubmarine patrols. SORG believed that a careful analysis of tactics would help determine which ones were effective and which ones were not. Voge assumed that if there were any glaring flaws in the plans for Operation Barney, SORG would find them and offer suggestions on how to fix them. In addition, Voge knew that SORG would offer an opinion on the mission's chances of success and its likely outcome. Lockwood had concurred in Voge's decision to seek SORG's advice. He also had no reason to doubt that Operation Barney would receive the panel's unequivocal support.

On April 18, Lockwood hopped a flight from Guam to Pearl Harbor to wrap up loose ends pertaining to Operation Barney, after which he and selected SubPac staffers would fly to San Diego for the mine-detector competition. Despite the long flight to Hawaii and with only a few hours' sleep, Lockwood went straight aboard the *Skate* to witness tests of her FM sonar, and UCDWR's pillenwerfer decoy system, the Alka-Seltzer–like noisemakers that masked enemy sonar. Several subs, including the *Seahorse*, had been similarly equipped with pillenwerfers, and this was Lockwood's first opportunity to see one at work.

With the tests concluded, Lockwood pronounced the *Skate*'s FMS ready for service and Lynch's handling of it superb. The pillenwerfers, too. Departing for the States, Lockwood felt certain that Operation Barney would be a total success. What he didn't know was that while he and his colleagues were airborne over the Pacific, a Japanese hunter-killer group working around the Goto Retto had sprung a trap on the reconnoitering *Seahorse* that had Harry Greer and his battle-hardened crew fighting for their lives.

Days before the *Seahorse* was jumped by the Japanese hunter-killer group, she had conducted a recon of minefields off the east coast of Formosa. While reconnoitering she encountered a problem with one of her mine-clearing cables that had been hastily installed at Guam and now had parted. It had taken a heroic effort on the part of the *Seahorse*'s crew, working in frigid and rough water, to clear it.[1] The episode foreshadowed problems with mine-clearing cables that would later plague more than one submarine during Operation Barney.

Greer was also plagued by the Kuroshio Current, which set the *Seahorse* to the north at a two-knot clip during the period of her submerged mine recon. It took superb seamanship to prevent the current from twisting the ship sideways into the forest of cables tethering the mines. Greer knew that the Kuroshio Current would have a similar effect on subs up in the Tsushima Strait and made note of its strength in his patrol report.[2]

Greer radioed Pearl with the results: ninety-seven densely packed mines in a field just north of Formosa, accounted for and mapped. With his mission accomplished, Greer made tracks for his next patrol station outside the Tsushima Strait, where the *Bonefish* would later conduct her lifeguard assignment.

The *Seahorse* arrived in the area on April 9 and within hours Greer heard distant exploding depth charges. There was, he noted in his patrol report, a heavy presence of enemy antisubmarine patrols.

0800 Hear two "pingers" [echo-ranging] far away but getting nearer.

1022 "Pingers" steadily coming closer. Evidently making a sound search off southern and eastern end of TSUSHIMA. Rigged for depth charge and silent running.

1130 Current setting us on to TSUSHIMA at rate of one
to one and one-half knots and unable to get out of the
"pingers'" road. Decided to sit on the bottom and wait
it out.

1220 Pingers" passed overhead—one down port side
and other down the starboard. . . . [3]

Days later, Greer tangled with two crafty patrol boat skippers south
of the Tsushima Strait.

First team patrol boats are used to patrol south of TSU-
SHIMA. They are equipped with radar similar to our SJ. . . .
Also suspect they carried radar detectors. They seem to patrol
in pairs with usually a radar-equipped plane searching with
them. The planes are much in evidence during the nights. The
patrol boats are equipped with large depth charges which do
not make the usual detonator click before exploding.

The Straits of TSUSHIMA live up to their reputation of
being very strongly defended. There is every evidence the
enemy is making maximum use of radar and radar detection
in his defense. Our SJ, which has given us such an enormous
advantage, is being made almost a hindrance by [this]. . . . The
enemy's . . . anti-submarine measures were so constantly active as to give us the feeling of being on the defensive instead
of the offensive.[4]

Japanese patrols dogged the *Seahorse* for days as she shadowed potential targets, which, like galloping ghosts, faded away into misty coastal
shallows or at times, it seemed, into thin air. Greer parleyed with Edge
early in the morning on the sixteenth, when the *Seahorse's* appearance
proved that she hadn't been sunk. Greer warned of the heavy coverage by
radar-equipped patrols sniffing around the strait. Edge acknowledged the
warning, then departed for the *Bonefish's* assigned lifeguard station in
those very waters. Greer headed off to carry out a mine recon of the

strait's southern approaches. His experience with patrol boats gave him good reason to believe that the Japanese knew the *Seahorse,* if not the *Bonefish,* was patrolling south of the strait. "The Japs probably wondered why I didn't go in or go away," he said.[5]

Midday on the seventeenth, Greer saw a telltale wobble of interference shimmer across the *Seahorse's* SJ radar screen. He assumed it was coming from a friendly SJ radar, most likely the *Bonefish's* or *Crevalle's,* though a challenge keyed from the *Seahorse's* SJ went unanswered. Greer ignored the interference and moved on. Two hours later another wave of interference, stronger than the one before, ghosted across the *Seahorse's* radar screen. It should have warned Greer that he was sailing into a trap set by the Japanese; somehow, it didn't. After the *Seahorse* submerged at daylight, Greer's sonar watch even reported intermittent pinging on the same bearing as the interference, as well as distant explosions, possibly depth charges, a sure sign that patrol boats were in the area. Back on the surface late in the day, the *Seahorse* picked up more SJ interference, which this time Greer attributed to a radar-equipped escort.

A day later the *Seahorse* picked up yet more radar interference. Despite the known presence of radar-equipped patrols, for some reason Greer thought the interference was coming from the *Crevalle.*[*] Almost too late, Greer discovered that he'd made a big, possibly fatal mistake. As he later said, "They were Japs."

18 April

0512 Radar contact on two small targets at 8,000 yards [roughly four miles]. Radar interference very strong and steadied on us.

0514 Dawn is breaking and sighted two patrol boats— very hazy. Opened out at full speed. Escorts appeared

---

[*]Why Greer believed this to be so isn't made clear in his patrol report. Submarine radar operators were highly experienced in such matters and likely the radar watch aboard the *Seahorse* made an interpretation that Greer must have concurred with.

**larger than PCs [patrol craft] but not as large as
destroyers. Range opened out slowly as smoke poured
from the escorts.**

The *Seahorse* hauled out on four mains with water spouts from the escorts' guns rising in her wake and then all around her.

Greer pulled the plug, ordered the *Seahorse* to three hundred feet, rigged for silent running and depth charge. He changed course ninety degrees to get off the escorts' inbound track, then fired two pillenwerfers to confuse them. A half minute later the pinging patrol boats passed astern, only to come about and launch an accurately aimed salvo of nine depth charges that exploded above and on either side of the fleeing submarine.

The salvo slammed the *Seahorse* down to four hundred feet as if she were a toy submarine, not one made of fifteen hundred tons of steel. Greer stopped the motors and put her on the bottom to wait out the attack and to take stock of the damage the initial depth charging had caused. It was bad.

Seawater pouring into the control room through the SD radar mast's shattered packing gland, and into the forward engine room through the warped main air induction valve endangered the submarine. Ruptured main fuel ballast tank vent fittings and hydraulic lines had sprayed diesel oil and hydraulic oil into damaged compartments, their noxious odors permeating the ship. Acid slopping from cracked battery cells threatened to form deadly chlorine gas. Cork insulation knocked from bulkheads and glass from shattered lightbulbs littered the decks, crunching underfoot. The most serious damage was yet to be discovered as men went to work in the eerie silence of a dark, half-dead ship, groping to make repairs to smashed equipment by the light of portable battle lanterns.

"All in all," said Greer, "things looked pretty grim. The ship was a shambles. Another pattern of depth charges like the one already delivered could possibly have finished us."[6] When he said that, he didn't know that a crack had opened around the ship's steering wheel mounted on the pressure hull in the conning tower. Given the submarine's great submerged depth and the enormous weight of water pressing against her, the crack could lead to hull collapse. A thick-skinned *Balao*-class submarine, the *Seahorse* was nevertheless bottomed in mud below her test depth of four hundred feet. A depth charge exploding close aboard in an uncompressible medium like seawater could easily finish her off.

Around dawn, sonar reported the arrival of a third pinging patrol boat that delivered two depth charges that drove the *Seahorse* deeper into the mud. After this cracking bombardment the three patrol boats drifted away, their screw noises fading. Greer and his crew waited in silence, all movement, all repair work suspended. Time passed slowly, the seconds ticked off by the ship's chronometers, those that still worked, and by the steady drip, drip, drip of water into the bilges from leaking sea valves, cracked freshwater engine cooling lines, and packing glands. The ship's breathable air, already thick with oil fumes and carbon dioxide, turned foul: eighty-plus men consume a lot of oxygen and respire a lot of CO_2. Greer wasn't fooled into taking action by the patrol boats' disappearing act. He knew that they could be up there lying to, engines silent, ready to bore in with their guns and depth charges, just waiting for the *Seahorse* to stir from the muck and poke up a periscope.

When Greer thought it was safe to move around inside the ship, he gave the okay to restart repairs, but quietly, and he cautioned the men not to drop any tools. Meanwhile, a stem-to-stern survey disclosed yet more damage: The reduction gear lube oil cooler had ruptured but was repairable; the air-conditioning system's Freon lines had ruptured, too—Freon gas is as deadly as CO_2; both periscopes were flooded and useless; the gyro compass was out of commission; both service radio transmitters were dead, their topside antennas and grounds carried away by the force of exploding depth charges; the ship's food storeroom had flooded. And on it went. The forward and after torpedo tubes and their outer door interlocks had jammed; all four inboard engine exhaust valves leaked seawater into the engine rooms.

When sonar reported the buzz of fast screws a few miles north of the *Seahorse*'s position, men froze in their tracks; repairs came to an abrupt halt. Greer could tell from the screws' fast-changing bearing rates that the patrol boats had lost the scent and were searching for the *Seahorse* inside a large circle several miles away from where she lay. As the sonarman, drenched with sweat and laboring for breath in the fetid atmosphere, listened to the patrol boats make a circuit, he heard them drop almost two dozen depth charges. Even though the drop was far off the mark, its thunder rattled the *Seahorse* and shook up her weary crew. How close would the escorts and their depth charges come? they asked. The men waited expectantly, until the thunder slowly faded away to ringing silence. Minutes passed. When it was all clear, the men resumed their work.

An exhausted Greer toured the ship to see for himself what repairs had been made. What he found was nothing short of a miracle, given the conditions the men had to deal with. They had done everything in their power to restore, repair, and jury-rig the ship's systems to get her off the bottom and back to the surface. As Greer said later, "[The] men were not defeated, on the contrary they were just beginning a long and successful fight." Back on the surface they could start for home—if the Japanese didn't return in force.

Toward midnight Greer made a circle in the air with an index finger, indicating a 360-degree sweep on the listening gear. Had the Japanese departed or were they still up there, waiting? After completing a slow, careful sweep, the sonarman gave Greer a vigorous thumbs-up—*All quiet, Captain.*

But would the leaking batteries have enough current to drive the motors and turn the props? Would the props even turn, or were their bearing collars misaligned and jammed? Was there sufficient compressed air in the high-pressure bottles to blow the main ballast tanks dry and lift the partially flooded sub from the bottom? The *Seahorse* was saddled with tons of water that couldn't be pumped overboard until the noisy trim and drain pumps became fully operational.

Greer looked around at the sweat- and oil-burnished faces of his exhausted crew, men who had given the *Seahorse* everything they had to save her and themselves. It was time to put their courage to the test yet again. He gave the order to blow the ballast tanks, to get the ship up and off the bottom.

High-pressure air roared into flooded tanks. For a minute the *Sea-horse* refused to budge, to acknowledge that the time had come to rise from the muddy sea bottom and shake herself free. Then, slowly, almost imperceptibly, she lurched up bow first. Frames and hull groaning in protest, she rose to a chorus of cheers from her crew. They were going to live! They would breathe fresh air again! They would get the rusty old bitch back to Guam in one piece after all!

It seemed a lifetime before the *Seahorse* reached the surface. Her one working depth gauge still registered seventy feet, but seas breaking over her main deck as she rolled and pitched like a surface ship on long ocean swells said that she was there. The bridge hatch and main induction clanged open. A hurricane of fresh air so sweet that it was almost intoxicating swept through the ship, purging the poisonous atmosphere and filling the men's lungs.

Motor macs started the engines and ran them up. The starboard reduction gear was far too noisy but turned its propeller shaft. SD radar was a wreck, flooded just like both periscopes. Somehow SJ radar, with its delicate innards, had survived the depth charging. Switched on, it immediately picked up the now-dreaded interference, a sure sign that radar-equipped Japanese patrol boats, maybe even the same ones that had almost killed the *Seahorse*, were somewhere in the area.

Greer hugged the western coast of Kyushu, limping along at the best speed his damaged sub could manage. He was careful not to sweep for targets with SJ, using it only to register bearings on the interference, which grew weaker as the *Seahorse* headed south, hugging Kyushu for cover. A little past dawn of the twenty-first Greer submerged "gingerly," as he put it, to test the ship and to make more repairs, some of which required welding cracks in the main engine-cooling system.

Heading southeast, away from the East China Sea, making for Guam, the realization dawned on Greer that the damage to his ship would require major work to repair: work only a shipyard like the one at Guam could provide. Clearly the *Seahorse* could not be repaired in time to join Operation Barney. Another sub would have to take her place. Preoccupied with his current plight, Harry Greer didn't have time to worry about Law-rence Edge's *Bonefish* running into radar-equipped patrol boats. In fact, Edge was busy trying not to run afoul of mines.

While the *Seahorse* limped home, the *Bonefish*, submerged west of Danjo Gunto, began her approach on the northeastern end of the line of mines Edge intended to survey. Almost immediately hell's bells, as if trying to clear their throats, issued a garbled bell tone. A few dim and shapeless green blobs appeared on the PPI scope, only to fade out. Edge wasn't fooled. The contacts' poor shapes, poor tones, low persistence, and erratic movement told him that FMS had homed in on a school of fish, not mines.

The *Bonefish* pressed on, her crew tense—these were not friendly wa-ters and the mines they were hunting weren't dummies. Edge, too, wiped sweaty palms on his khaki trouser legs. Then: "Mine contact!" The first strong, clear-toned contact of the morning. Range 250 yards. Then an-other, almost dead ahead. Other mines on the line began to show up at regularly spaced intervals and then all along the same line where the first two had been contacted. Hell's bells' clear, sharp ringing tone and the distinct, bright, and evenly spaced green pears on the PPI left no doubt that these contacts were Japanese mines, not fish. In his top secret report to Lockwood, Edge noted, "The tones were so clear and bell-like that all other contacts, of which there were many, could not compare and caused no confusion."[7]

Edge experimented with the FMS, varying the ship's speed to see if it had any effect on the contacts' range or tone. It didn't. For the remainder of the day Edge ran the *Bonefish* through gaps in the line and along the line itself, plotting the layout of the single and double rows.

Edge tried to determine the depths at which the mines had been planted, but couldn't establish anything definite. He upped the scope while submerged to get a glimpse of the mines, trying to determine their depth, but the murky water limited his visibility. Aside from that, the *Bonefish* was never closer than two hundred feet from the line. Varying the sub's depth changed the size of the blobs on the PPI, but not their tone, range, or clarity. Based on this experiment, Edge believed that most of the mines had been planted shallowly.

Confronted with data that seemed to confirm this belief, Edge casually commented in his report, "It is also interesting to note that Bonefish, in proceeding to assigned area . . . on 12 April, 1945, while on the surface, crossed this same mine line at about 19 miles from its NE end without incident and with beautiful bliss."

Plotting resumed the next day. Edge concluded that the minefield was forty-seven miles long, with fifteen mines per mile, and that its double rows were 150 yards apart. The exercise revealed how accurate FMS plotting could be in the hands of expert operators and how its sensitivity and flawless performance could instill a feeling of confidence in a submarine's crew.

The data collected by Edge and the other skippers of FMS subs sent north by Lockwood, along with the intelligence developed by ICPOA, would provide the Operation Barney submarines with, if not a perfect picture of what lay hidden in the Tsushima Strait, a picture that would prove to be remarkably accurate.

Mine recon completed, Edge moved north to patrol near Quelpart. Early in the morning, tracking what he thought were luggers, Edge almost stumbled into a trap set by two destroyer escorts. The DEs had the *Bonefish* boxed in between themselves and a pair of islands off the coast of Kyushu. Rather than tangle with these ships, Edge pulled away at flank speed to outrun them. The DEs were fast, faster than the *Bonefish*, and, closing in, they opened fire, forcing her down.

> Submerged with the first shell passing overhead, and splashing about 200 yards over; range 4100 yards!
>
> [Minutes later] Seventeen depth charges, all uncomfortably close, went off as we doubled back under the DEs at 270 feet. We had to turn under them due to the proximity of land in the other direction, and it may have been this which fooled them since screws passed overhead about two minutes before the first explosion.[8]

From early morning until early afternoon, Edge heard strong echo-ranging and counted

> . . . a total of 41 more depth charges dropped as if the boys were hot on something's trail; but this time it was not us! Total for the day was 75 depth charges [the last of which despite being far away] rattled the ship—possibly due to shallow water.
>
> Surfaced, very much put out by the day's experience.

Two days later the *Bonefish* came close to another run-in with one of the depth-charge-dropping DEs. This time Edge kept his distance, remarking that their "treatment at his hands was too fresh for us to feel like playing with him again."

Harry Greer's warning about near-saturation antisubmarine patrols had proven correct. In the closing days of the patrol Edge recorded fifty-five contacts with Japanese patrol boats, escorts, and planes in the lower Tsushima Strait area. Later, endorsing Edge's patrol report, Division Commander Louis Chappell said it all.

> Enemy anti-submarine measures were intense, persistent and effective in that it prevented [the *Bonefish*] from closing the Japanese coast and kept her on the defensive a good part of the time. The Division Commander concurs in the opinion that there is a strong probability of an integrated effort involving enemy air and surface craft and shore based radar stations [to thwart submarines].[9]

———

Edge at last found time to work on the letter to Sarah he had started weeks earlier.

> *All I'll say . . . precious, is that I love you intensely today and forever.*
>
> *The patrol goes on but, but still no good luck [finding targets], except that part of our special mission has been*

completed successfully. I say "no luck," but at least the boat is still safe; so the luck isn't yet completely ill. And there are yet a few days even if experience so far doesn't make them promising. . . .

Darling sweetheart, how I miss you! And how much I am wont to dream about you, and dream about our being together again, in our own home, and what we'll do and feel like again. It's a very, very wonderful dream and even if in reality when that time does come, it is only half as wonderful as the dream, it will still be wonderful.

Hitler is dead!* I do hope so! I don't care how he died, just so long as he is really dead. I have no interest in his dying a death of torture, because he could not be tortured enough . . . to pay for the torture of [those] he and his henchmen brought to so many millions of people. . . . At least we can say there is a vast difference in his death and in that of F.D.R. I wish I could have more faith in the results of the San Francisco Conference [to establish the United Nations], or any world peace movements. . . . Not that I'm not all for trying to gain it in every way possible—as long as we don't lull ourselves into too great a disarmament too soon.

Well, this patrol is rapidly drawing to a close, and not too soon for any of us, either. . . . Anyhow it's going to be very, very good to get in—and best of all receive your letters, my dearest wife.

I know you were probably disappointed by my last letter before departing, the one in which I told you the admiral himself told me this was not to be my last patrol after all. . . . I still don't like these things and it will be a relief to feel that the

---

*Fleet Radio and the Armed Forces Radio Network announced the news the day after Hitler committed suicide in his bunker on April 30, 1945. According to the journalist William L. Shirer, the announcement of the Führer's death came on Hamburg radio. It was preceded by three drumrolls followed by, "*Achtung! Achtung!* Adolf Hitler, fighting to the last breath, fell for Germany in his operational headquarters in the Reich Chancellery." No word that he'd blown his brains out.

bosses feel I've done my share of it, or at least can now serve in as useful a capacity elsewhere. . . . Anyhow, I guess I'm good for at least one more, which is what he [Lockwood] said when we left. Maybe it can be a good one, although from now on, I fear my main *worry will still be to get back safely—which is not the way heroes are made. . . .[10]*

Now that we are actually out of the area and well on the road to the base, I feel slightly like a different person already! Every minute is busy these days, though getting all the various reports together, cleaning the ship, qualifying new boys [in submarines], giving exams, making track charts, etc. Nearly everyone is busier than while actually on patrol. We were told to hurry back, and that is what we are doing, with the result that there is less time than usual to do all these things. [When I get back to Guam] what I really want is you. But maybe it's better that you won't be there, because I'd probably find it impossible to leave on another patrol . . . !

Bye for a little while, my loveliest one; I'll be writing again soon. . . . With all my deepest, dearest love and warmest kisses,

Lawrence[11]

In San Diego, Lockwood had received reports on the *Seahorse*'s excellent mine recon and near-fatal run-in with Japanese patrol boats. Then from the *Bonefish* came Edge's radio message outlining the stunning success he'd had plotting mines off Danjo Gunto. Lockwood was flush with pride over the two skippers' work. He was proud, too, that he'd not wavered in his belief that FMS held the key to Operation Barney's success and that it was the best of the competing sonar systems he and his colleagues had come to the West Coast to evaluate. Even the *Seahorse*'s plight—he was relieved that she'd made it back at all and impressed with the performance turned in by Greer and his crew (there'd be medals for everyone)—didn't slow his drive to get the operation under way on time. Another sub, the USS *Sea Dog* (SS-401), and her fine skipper, now on their way to Guam from a war patrol, would have to take the *Seahorse*'s place.

Despite all the unfinished business awaiting his return to Guam, Lockwood relished the three days he'd spent at sea testing sonar systems aboard the *Flying Fish* with her skipper, Robert D. Risser of Chariton, Iowa, and aboard the *Redfin* with its CO, Commander Charles K. Miller. Lockwood had been joined by Harnwell and Henderson, as well as officers, scientists, and electronics specialists from other research facilities and Navy bureaus. Lockwood had also brought along George Pierce of the *Tunny*, who, with his firsthand experience running minefields, could explain the practical side of the business.

The tests conducted off the coast of California were exhaustive and exciting for naval officers and landsmen alike; it was an adventure in submarining and mine-detecting procedures some had never had before and might never have again. The tests proved conclusively that FMS worked better than the other competing systems, which had major drawbacks in their operation, display, maintenance, and especially sensitivity. In the end, Lockwood had his way: FMS was what he wanted and what he got. There was no turning back; he'd staked everything on FMS and it had come through for him when it had to.

In an interesting footnote to these tests, Frank Watkins in Washington sent a letter to Lockwood that must have pleased the admiral no end.

"SORG has submitted a study of British Mine-Detection results to Cominch," wrote Watkins. "This study reputedly shows that British subs can do a fine job of mine detection. They conclude the study with a recommendation that British subs run interference for you thru Tsushima to chart the fields.

"Their study is based upon operational reports which show mine detection successes. Unfortunately, they do not have reports of the <u>unsuccessful</u> runs."

Before returning to Pearl Harbor and then Guam, Lockwood told Risser, CO of the *Flying Fish*, to unload the test gear and swap it for FMS, then get his sub out to Guam as soon as possible. It was already May and Risser's crew would need all the training they could handle; the other subs

Lockwood had selected to take part in Barney were beginning to assemble in Guam.

Flying west from California, Lockwood consulted a list of his Barney skippers, looking for the man among them who could lead the operation. Every skipper on the list had the necessary experience. They had also proven that they wouldn't buckle under pressure. They were an aggressive and tenacious lot, each man equipped with a sub commander's most valuable asset: brass balls. Lockwood may have had in mind a man like Dudley Morton of *Wahoo* fame, but without his dash and his propensity for taking risks. The man he was looking for had to have something special: maturity. Running a finger over his list, Lockwood came to a name, stopped, and put a check mark next to it. He had his man.

CHAPTER TWELVE
"Hydeman's Hellcats"

BULLETIN by Edward Kennedy[*]

(Associated Press Correspondent)

REIMS, France, May 7, 1945—Germany surrendered uncon-
ditionally to the Western Allies and the Soviet Union at 2:41
A.M. French time today. The surrender took place in a little
red schoolhouse that is the headquarters of Gen. Dwight D.
Eisenhower.

A t Pearl Harbor the surrender of Nazi Germany added a sense of
urgency to the launching of Operation Barney. Lockwood was impatient to finish the job U.S. subs had started, and to do it without
help from Halsey and Spruance, much less the Russians. The possibility
that the USSR would declare war on Japan, once thought remote, might
happen any day. That country's entry into the Pacific war increased the
likelihood that the Sea of Japan would have to be divided up into zones of
operation for both the United States and the Soviets. It also increased the
possibility that a Soviet Far Eastern offensive would overrun Manchuria,
even Korea. If that happened, Stalin would establish a Soviet-dominated

---

[*]The AP flashed the news a day ahead of the official announcement by the Allies. Kennedy, who jumped the gun on the bulletin, was suspended by Eisenhower's headquarters.

sphere of influence as well as military bases in a region the United States expected to control. Worse yet, Stalin had the will and the necessary forces to launch an invasion of Japan through Hokkaido, from staging areas on Sakhalin Island. For now the attitude in Washington was "wait and see."

Lockwood balked at the idea of Soviet naval forces, particularly Soviet submarines with little or no experience with war patrols, operating in the Sea of Japan. He feared that Soviet submarines would further complicate the already thorny business of identification of ships and units, which in turn could lead to disastrous consequences for U.S. sub crews, never mind Soviet ones. The only assistance Lockwood wanted from the Russians was an agreement that if an Operation Barney sub became disabled, it could seek refuge in Vladivostok. Otherwise Lockwood had no use for the Russians. Nimitz, despite his earlier aversion to this plan, was again making sounds that Lockwood should start thinking about how to divide up the Sea of Japan into U.S. and Soviet areas. Lockwood hoped that if he dragged his feet long enough, the whole stinking idea might be dropped.

The *Bonefish* arrived in Guam the day Nazi Germany surrendered. The long-expected news of Germany's defeat, when it arrived, rather than inviting celebration among all the military personnel on Guam, evoked a mood of somber restraint. Welcome as the news was, every American soldier, sailor, marine, and airman, from the top brass down, knew that the war was only half won, and what it might take to end it.

This was the underlying mood as the *Bonefish* tied up at the sub base in Guam. Lockwood was not there to greet her, as was his custom when subs returned from patrols, especially one as important as the *Bonefish*'s had been; he was still in San Diego and wouldn't return until the eighteenth. Barney Sieglaff, covering for Lockwood, and accompanied by staff and a Navy still photographer to record the *Bonefish*'s homecoming, came aboard to greet Edge and his crew. As Edge, Sieglaff, and the other officers gathered in the sub's wardroom over coffee and cigarettes, the two Japanese prisoners Edge had captured were escorted topside for transfer into the custody of waiting U.S. Marines. Suddenly, in an unguarded moment, one of the prisoners, the quiet one who had not begged Edge to kill him, bolted. Diving over the side of the sub into the water, he quickly swam away from the ship and drowned himself. The Marines and sailors who witnessed this event were stunned. A small boat put out after the prisoner, but it was too late.

Ashore after all the commotion had settled down, Edge found a pack of letters waiting for him, which he hungrily dug into for news about Sarah and Boo. In an earlier letter, one that arrived before his departure on the mine recon, she told him that Mack Tharpe, the husband of their close friend, Jane Tharpe, one of Sarah's bridesmaids, had been killed aboard an aircraft carrier. Even before answering Sarah's letters, Lawrence wrote a letter of condolence to Jane. His remarks shape a perfect coda to his own life: He had no way of knowing that he and Mack would share the same fate.

> *Dear Jane,*
>
> *My boat and I have just recently returned from another war patrol, and I am hastening to write now before the press of events again becomes too strong for me to compose my thoughts well enough to write coherently. . . . Words cannot possibly undo . . . the pain.*
>
> *I guess all of us out here give some thought to dying, and we all feel, I think, that if the war brings it to us that, even though we do want to live out a long life as much as anyone, leaving early in a cause which is after all, to us individually, one which is primarily to secure for you, our wives and kids (and their kids), the kind of life we've all known and [loved] in the U.S. [our] whole lives. If that cause should bring our death, then to say the least it is just about the best cause for which one can die young.*
>
> *Mack must have felt that way, I know. He didn't have to go to war, as he did. . . . Therein lies the pity of it—but therein also lies the greater honor of it, that which made his sacrifice of*

the very highest. I was always proud of knowing Mack. I was
never so proud as now.

It is the unfulfilled promise for the future years of life . . .
which makes it hard to bear the loss of a young man's life.
Mack's life was certainly one of those filled with all the brightest
of promises. . . . To you who loved him and gave him a family it
is more than that, I well know. It is part of your own life that
went with him, and the sacrifice is no less yours. I hope time
will make it easier for you to bear, even though the loss itself
can not possibly grow less. I hope the baby will make it easier,
too. Your courage in starting your family in the midst of the war
will give you that reward, I am somehow sure.

In as much as Mack died for you and the baby, so did he
also die for my two Sarahs, and for all of us because we all reap
the benefit of his sacrifice. So I humbly thank God for Mack and
the thousands of others who have died in this war. They
perhaps even more than the living make our lives and our
country what they are and what they will be.

Very sincerely,
Lawrence[1]

Writing to Sarah, Lawrence told her about the suicide of the Japanese
prisoner he'd brought to Guam for interrogation. In another letter he told
her that he'd

Received the unwelcome news today that an official
investigation is being made of the escape of our prisoner that
day we got in. As a result I've got to spend tomorrow . . . on our
tender at the hearing, darn it. Next time I may not bother to
pick up any darn prisoners. . . .[2]

Edge and his crew transferred temporarily to the submarine rest camp
for a few days of relaxation before returning to sea for training and FMS

trials. They knew that they had been tapped to become part of a special mission everyone was openly referring to as Operation Barney, and that it would entail dodging minefields and penetrating the Sea of Japan. The details of the operation were still under wraps. Even so, Edge knew they'd soon be on the table. Scuttlebutt had it that Lockwood planned to conduct a comprehensive mission briefing and that the initial jump-off date for the first group of subs had been scheduled for May 27.

With the clock counting down to that jump-off, Edge wrote to Sarah during the little spare time he had, given the demands that training and preparation for Barney had imposed.

> *Dearest Angel,*
>
> *Out here the only news is that we've been back on the boat now for several days of training. The period at [rest] camp certainly seemed short, but it couldn't be helped, and even full periods always seem short too; so I guess it really makes no difference anyhow. Wish someway this next patrol could be a short one too, but I don't foresee it. Certainly, it is scheduled to be a regular full-time one. . . . At the moment, I really feel more like going on patrol this time than I did last time, maybe because I had been away from it long enough to lose a bit of confidence— although I think I can truthfully say that I've never been over confident on the eve of departure.*
>
> *. . . I love you and miss you almost too much to bear at times, but I'm just living for the time when I can return to you and Boo. . . .*
>
> *With all my dearest love,*
> *Lawrence*[3]

Lockwood returned to Guam from San Diego. Aboard his flagship in Guam, the old sub tender USS *Holland* (AS-3), he looked out over Apra Harbor at the tender USS *Apollo* (AS-25), her fleet of nine FMS subs nested alongside. Lockwood had reason to be pleased. All of his and his staff's hard work was about to pay off. The plan they'd developed for the execution of Operation Barney was ready for presentation to the nine skippers. The increased tempo of FMS training had resulted in a marked improvement in the sub crews' attitude toward FMS. They might not all be converts, but even so, requests for transfers had petered out. As for FMS itself, the *Flying Fish* had turned in some of the best results that Lockwood and Sieglaff had ever witnessed, better even than the *Bonefish*. Robert Risser, her CO, had no combat experience with FMS, but was eager to have the opportunity to use it to get at those ships in the Sea of Japan. Lockwood noted with satisfaction that Risser's buoyant enthusiasm had infected the other skippers.

One of them was Commander Earl T. Hydeman, skipper of the *Sea Dog*, Lockwood's choice for the job of task force commander of Operation Barney. Hydeman's sub had been outfitted with the FM sonar suite removed from the damaged *Seahorse*, now undergoing battle-damage repairs in one of Guam's big floating dry docks. Lockwood had announced the skipper's selection upon returning to Guam from the West Coast sonar competition. Sieglaff and Voge knew Hydeman well and heartily concurred in their boss's decision.

Lockwood could not have made a better pick. A Midwesterner from St. Louis, Missouri, the thirty-four-year-old Hydeman had graduated from the United States Naval Academy in 1932. After a typical tour of duty aboard a battleship, he reported to submarine school for instruction, then for duty aboard the *S-18*. In the mid-1930s, Hydeman, like most young submarine officers, had tours of duty in a variety of subs, both S-boats and fleet boats, including Lawrence Edge's first sub, the *Narwhal*. After a stint aboard a destroyer and then a detour ashore at BuPers, by 1944 Hydeman itched for a command of his own. To achieve his goal he spent six weeks at PCO school, after which he completed a tour as a PCO aboard the USS *Pampanito* (SS-383). He then took command of the *Sea Dog* after she had completed her second war patrol out of Pearl Harbor under Commander Vernon L. Lowrance. Hydeman skippered the *Sea Dog* on her third patrol in the Nampo-Shoto area and the east coast of Kyushu in command of a three-sub wolf pack he named "Earl's Eliminators." Despite having sunk only one Japanese ship, a big

6,800-ton freighter, Hydeman turned in an aggressive and well-conducted patrol.

In making his decision to name Hydeman as Operation Barney's overall commander, Lockwood weighed several important factors in his favor. Hydeman, the senior CO in the group, had proven by his record of achievement that he possessed the requisite maturity, experience, and sound judgment. He also possessed something else, something special that the leader of this mission would need, and which Hydeman had in abundance: iron nerve.

Having spent a good part of the war in New London and Washington, D.C., Hydeman had had little combat experience, far less, in fact, than some of the other Hellcat skippers he would command. But he knew submariners—what made them tick and what made them successful. He'd gained this knowledge as personnel officer for ComSubLant, where, among other things, he detailed enlisted crews to new-construction subs. Therefore he understood the importance of having highly qualified sailors, not just volunteers right out of boot camp, to man the boats. Hydeman was a crusader on matters of personnel selection and training. Lockwood knew what Hydeman had accomplished and the benefits of it that had accrued to the sub force, and to the FMS sub crews themselves.[4]

Hydeman, like Lockwood, never doubted that Operation Barney would succeed. The only concern he had, which he expressed to Lockwood and Sieglaff, was that his and the Sea Dog's arrival had been a last-minute affair and that the Sea Dog's substitution for the Seahorse hadn't allowed much time to train on FMS and to get the hang of locating mines. Sieglaff waved Hydeman's concerns away, saying that Hydeman's crew would get the hang of it in no time, which in fact they did, despite numerous equipment failures and the corresponding repairs that often lasted until dawn.

"Well, Hydeman," Lockwood drawled during one of their frequent meetings, "what're y'all going to call this little group of yours?"

Hydeman thought a moment. "How about 'Hydeman's Hellcats'?"

And so "Hydeman's Hellcats" became the official name attached to the Japan Sea Patrol Group. It was an apt name for the raiders. After all, a hellcat exists solely to torment others, in this case the Japanese, who, like the Americans at Pearl Harbor, were about to be caught by surprise. For purposes of further identification, Hydeman gave each of the three task groups its own name.

HYDEMAN'S HEPCATS
Sea Dog—Hydeman in overall command
Crevalle—Steinmetz
Spadefish—Germershausen

PIERCE'S POLECATS
Tunny—Pierce in command
Skate—Lynch
Bonefish—Edge

RISSER'S BOBCATS
Flying Fish—Risser in command
Bowfin—Tyree
Tinosa—Latham

That settled, Hydeman and Sieglaff pored over the mission's operational details, Hydeman offering valuable opinions and ideas on how to improve it. Hydeman then met with the other Hellcat skippers to discuss any concerns they had. The first group of three subs, Hydeman's Hepcats, were scheduled to sail on May 27. They'd be followed by the second and third groups on May 28 and 29, respectively. After the meeting Hydeman reported to Lockwood that his Hellcats were itching to go. To make certain that the mission got off to the proper start, Lockwood had arranged to lay on a lavish luncheon for the departing skippers and, to liven it up, invited as many Red Cross girls and nurses from the Navy base at Guam to attend as could be spared. Lockwood was indeed a big-picture man.

Between letters to Sarah, Lawrence kept his parents informed about his work, but only as much as he knew the censors would permit.

[O]ur period at rest camp is practically over again, and it won't be so long before we're on our way again. Yes, perhaps the submarines' part in the war is fast becoming much reduced. If this is actually going to be my last patrol, as I've been led to believe, I don't think I will be at all sorry. I'm getting a bit anxious to get back to some sort of radio or electronics job before I forget all I ever knew of it. I've even given some thought to trying to get into surface ships again, eventually, before I become too senior to get a good job in them. That, however, is looking a bit farther into the future than I need worry about now. First, I must at least get this next patrol over with successfully! And it will probably be a long one, at that, instead of short as was this one just completed.

. . . You asked whether we in subs have to go pretty close to Japan now. The answer is mostly yes, but actually they have been doing that almost since the war began, those assigned to that part of the Pacific, that is. Nowadays, there's not much left of the ocean for our exclusive use except the waters close to Japan, and even they are often shared with our planes these days. Guess by the time I get a different job in the war, I'll have covered a fairly good portion of the western Pacific. Still wish I could see some of the places to which we pass so close at times!

Your comments on F.D.R.'s death were much the same as those attributed to most of the people in the country by Time, whose issue covering the week of his death I read only last night. Guess we all feel similarly about it—and about Truman.

I'll write again soon. Meanwhile, worlds of love to each of you.

Affectionately,

Lawrence[5]

The Operation Barney briefing began as scheduled for the skippers and their communications officers aboard the *Holland*. Security was tight. Blackout restrictions required that all of the ship's portholes and watertight doors remained secured. It was a sweltering night and the *Holland* lacked air-conditioning. Voge, his khakis blackened with sweat, started off screening a film produced by UCDWR in San Diego under the direction of Harnwell and Henderson. The black-and-white film showed every phase of FMS operation, training, and maintenance. Close-ups of juicy green pears on a sonar screen and a sound track of tolling hell's bells completed the picture.

Voge then launched into a full review of Operation Barney, starting with its conceptual framework, moving on to the details of the Sea of Japan's hydrography and geography. He ended the review with an ominous warning: Stay out of waters mined by B-29s along the northwest coast of Honshu. Voge explained that the magnetically actuated bottom mines dropped by B-29s might not show up on FM sonar. These U.S.-mined waters were marked in red on the maps that had been provided for the conference. More information about them would be included in the skippers' op orders.

Voge reminded them, "If you get into such bad trouble that you can't make it out of the Sea of Japan, head for Vladivostok. There you'll have the status of a man-of-war of a belligerent nation entering a neutral port. Maybe you can effect repairs and get out in time. At the worst you'll be interned for the duration."[6] He told them that if this should happen, they were to contact the U.S. consul in Vladivostok to claim sanctuary. They were warned not to enter Soviet territorial waters except in an emergency. The Russians patrolled these waters, which were probably mined, to prevent incursions by foreign ships.

After Voge completed his presentation, Barney Sieglaff took over. He had prepared the tactical plan that divided the Hellcats into three groups of three subs. He had also prepared the departure schedule of the three task groups from Guam, the timing of each group's entry through the Tsushima Strait into the Sea of Japan, their individual area assignments, the date and timing of the Hellcats' initial attacks, and the date and timing of their departure through La Pérouse Strait. He'd also developed the wolf-pack-style codes for use by the Hellcats en route and during the operation itself.

Sieglaff gave the skippers a detailed description of the minefields located in the southern approaches to Tsushima, based primarily on the data gathered by the *Spadefish*, *Seahorse*, and *Crevalle*. Sieglaff then explained what ComSubPac knew about the actual minefields the raiders would encounter in the Tsushima Strait, based on intelligence gleaned from sources including prisoners, spies, decrypts, etc. The sources confirmed that three, possibly four strings had been sown across the western channel. A thousand-yard gap separated each string, with fifty yards separating each mine from its neighbor. The mines themselves were said to be sown at depths of twelve feet, forty-two feet, and seventy-five feet. According to Sieglaff, as far as ComSubPac could determine, no changes or additions had been made in the layout of the minefields since they had first been sown in late 1941, other than to replace breakaways caused by storms and the deterioration of anchor cables.

After departing Guam on their appointed days, the Hellcats would proceed independently in three groups at normal two-engine cruising speed. Sieglaff reviewed the codes that had been adapted from wolf pack operations for use by the Hellcats to communicate with ComSubPac and with one another during their voyage and during the raid. Sieglaff's codes had been developed to meet the need for a rapid exchange of information, and consisted of hundreds of stock phrases, each represented by a two-number code group for geographic locations or operational procedures. For example, "No. 24" meant, "Am heading for internment at Vladivostok." In addition, each sub had its own call numbers and letter groups to speed ship-to-ship communications and coordinate attacks. The *Bonefish* was 67V606.

As for the actual mine-penetration part of the mission, Sieglaff didn't mince words. "Inoperative FM sonar gear will not be considered sufficient cause to delay the transit of any ship. In case gear is not operative, the submarine will make transit at 200 feet or greater depth." In other words, go in naked like the death-defying Ray Bass in the *Plunger* when he ran through La Pérouse submerged back in '43. Sieglaff's words surely must have raised eyebrows aboard the *Holland*, for this went against Lockwood's dictum that going in under mines was a sure way to lose submarines.

Sieglaff had devised three governing days for the Sea of Japan raid:

Fox Day, Mike Day, and Sonar Day. Sieglaff had nominated Fox Day, June 4, for Hydeman's Hepcats' submerged transit through the western channel of the Tsushima Strait. Pierce's Polecats and Risser's Bobcats would make their transits on Fox Day plus one and Fox Day plus two, respectively. The nine submarines would begin attacking targets at sunset on Mike Day, June 9. The Hellcats then had fourteen days allotted to sink as many ships as they could, after which they would exit the Sea of Japan through La Pérouse Strait on Sonar Day, June 24.

Sieglaff stressed the importance of the subs remaining undetected in the Sea of Japan until Mike Day, when all the Hellcats would be in position. He warned them not to jump the gun by attacking too soon, no matter how tempting it might be, as attacking too early could jeopardize the mission by drawing Japanese antisubmarine forces to an individual sub. As it was, torpedo attacks erupting across the Sea of Japan would send enemy ships running for cover. Attacking too early would only make them run faster and make them harder to find. The only exception to this rule would be if an enemy man-of-war, say a heavy cruiser or carrier, were to show up before sunset on Mike Day. Sieglaff added that any submarine that expended all its torpedoes early should proceed to a position near La Pérouse Strait to await the arrival of the other Hellcats.

Sieglaff then addressed Sonar Day, June 24, the day of departure through La Pérouse. He had devised two exit plans, which he named "Sonar Xray" and "Sonar Yoke." Depending on conditions, the density of antisubmarine patrols, and the like, the Hellcats were to make their exit run either during the day submerged (Sonar Xray), or at night on the surface (Sonar Yoke). In the event that intelligence revealed increased antisubmarine activity in and around La Pérouse Strait, ComSubPac would radio a warning to Hydeman along with a recommendation as to which plan he should employ. Sieglaff stressed that the final decision would be left up to Hydeman. If he chose to use a submerged exit run, the Hellcats would go through La Pérouse in two parallel columns at three knots, picking up a knot or two of extra speed from the outflowing Kuroshio Current.

In regard to mines, there was little in the way of fresh intelligence to indicate that, like Tsushima, changes had been made to the fields presently in place or to the layout of the safe channel used by the Russians. Sieglaff stressed that if a submerged exit became necessary, any subs with inoperative FMS gear would have to try to make the run-through at a depth below the deepest-sown mines—whatever that was—and good luck.

If Hydeman decided to make a surface dash—Sonar Yoke—he was to keep in mind that there were shore-based enemy radar stations on Rebun Island west of Kyushu and at the naval station at Wakkanai Ko near the tip of Kyushu itself. Sieglaff warned that according to intelligence, the stations were fully capable of detecting ships and even low-lying submarines. If Hydeman chose the surface option, Sieglaff recommended making the dash through the safe channel at flank speed, all nine subs closed up in a single column, gun crews standing by to slug it out with Japanese patrol boats if necessary. Lockwood, as he had during the Japan Sea incursion of 1943, had arranged for a diversionary bombardment by submarine of an island near the Tsushima Strait to draw antisubmarine forces away from La Pérouse. In any event, after clearing the strait the Hellcats were to make tracks to Pearl Harbor, where Lockwood and his crew would be waiting to greet them.

Sieglaff and Voge answered a few questions from the skippers about the mission, then handed out copies of ComSubPac's top secret Operation Order No. 112-45, predated May 26, 1945.[7] Annex "A" of the op ord contained all of the information in Sieglaff's presentation for study and review. With that, Sieglaff wrapped up the conference.

Lockwood had remained silent during the presentations. Now he rose to address the Hellcats and give them his personal "blessing." The conference just ended, he said, had brought their plans to a climax. It was "the day we have lived for." He praised their professionalism, determination, and deep-down courage, adding that with their spirit, Operation Barney could not fail. Finally, he asked one favor of them: He had been a submariner since 1914 and had never fired a torpedo in anger. He asked that they fire plenty of them for him in the Sea of Japan, as he would give a "right hind leg to go with them on their mission." Then it was "God bless you and good hunting!"

After the meeting Lockwood returned to his stateroom to read a report he'd received earlier that day from SORG. It was a response to the Operation Barney prospectus that had been forwarded to SORG by Dick Voge for an evaluation. Late arriving, SORG's report was not what Lockwood had hoped for. SORG warned that Operation Barney entailed great hazards and that ComSubPac should expect heavy losses of men and ships. For SORG, the mission didn't appear to be justified by the probable damage, light damage at that, that the Hellcats could inflict on the Japanese. Those observations troubled Lockwood, but not nearly as much as SORG's comment that unless the strategic situation demanded its execution, *Operation Barney should not be launched* (emphasis in original).[8]

Lockwood trusted the judgment of the men who had made this evaluation. They were men he knew and respected, officer and civilian alike. They had vast experience in submarine tactics and operations and knew what worked and what didn't. They were fully cognizant of the losses so far sustained by the sub force and had worked hard to find ways to eliminate the gross dangers sub crews faced on patrols by helping develop tactics that would accomplish SubPac's goals while minimizing risks. SORG's board understood that war patrolling was not risk-free. But Operation Barney was more than just another war patrol and SORG knew it. Understandably, Lockwood was reluctant to disregard the advice and counsel SORG had offered. Sure, there was no way to know how many ships were still afloat in the Sea of Japan—intelligence estimates of their numbers varied wildly. And there was no way to know how many of them the Hellcats could sink. Yet whatever the number, it would weaken Japanese morale. There would be no way they could ignore the fact that their island fortress had been penetrated by submarines capable of destroying the emperor's lifeline to the Far East. As for the risks entailed in underrunning minefields, there was no question that there was a good chance one or more of the subs might be lost. Losing subs was a risk Lockwood had to take. If it happened he'd take full responsibility and accept the consequences, just as he had for each of the subs lost so far in the war.

Despite FM sonar's shortcomings and the risks posed by Operation Barney, Lockwood decided that he had to put SORG's conclusions aside. "Wars are not won that way," he said, meaning wars were not won by fearing to take risks when so much was on the line. That belief strengthened Lockwood's resolve as he contemplated the unknown. That and a voice calling to him: the voice of Mush Morton expressing thanks that Lockwood hadn't wavered in his hunger to exact revenge on the Japanese for the death of the *Wahoo*.

CHAPTER THIRTEEN
Running the Gauntlet

The Hellcats trained for their mission day and night right up to the date of departure. The exhausting schedule gave Edge very little time for letter writing. The only opportunity he had was late at night when the blackout on Guam and its strict enforcement forced him under a blanket with a flashlight. Somehow Lawrence found time to compose one last letter, which he dropped into the Pacific fleet mailbag on May 25, two days before shoving off on Operation Barney.

When it came time for the big send-off on May 27, Lockwood spared no effort to put his skippers in a relaxed mood. He had the *Holland*'s Filipino mess boys lay on a lavish luncheon buffet in one of the tender's spacious staterooms. As promised, a bevy of good-looking Red Cross girls and Navy nurses joined the party.

Lockwood felt that Operation Barney called for a send-off that would ease the strain and self-consciousness of bidding farewell and good shooting to the young Hellcat skippers. It also eased Lockwood's own mind, too, as the Hellcats prepared to depart. Notwithstanding his belief that SORG's estimate of losses was exaggerated, if not wrong, he still suffered deep misgivings and anxiety about the wisdom of sending men into the teeth of Japanese minefields with little to guide them but a fancy electronic device that wasn't always reliable.

Inevitably the clock ticked down; the party broke up. At three p.m., the traditional hour of departure on war patrols, Earl Hydeman indicated that he was ready to go. Diesels rumbling impatiently, the *Sea Dog, Crevalle,* and *Spadefish,* now moored alongside the *Holland,* cast off their lines and, with Lockwood and his people watching and waving good-bye, backed clear of the tender. Escorted by a destroyer, Hydeman's Hepcats departed Guam for empire waters.

Lockwood bade good luck and good hunting to Pierce's Polecats and Risser's Bobcats on May 28 and 29, respectively. As each group departed, he felt the familiar tug of apprehension and no small measure of doubt. But this was it; they had all departed, and Lockwood, stuck on the beach, could only wait impatiently for the first reports of action from the front lines to arrive after Mike Day, June 9, when the Hellcats were slated to begin their attacks.

Under way, the men aboard the nine Hellcat subs got about their business. The enlisted rates already knew what they were in for and needed only to have it confirmed by their COs, who a day out from Guam sketched in the broad outlines of the mission for their crews, filling in the details only as needed for the men to do their jobs.

Like all sub sailors, the Hellcat sailors were a cocky and profane lot— fatalistic, too: They were more than willing to look death in the eye so long as there was a reasonable chance they'd survive the encounter. At this stage most of them believed that their chances of surviving Operation Barney were better than even; otherwise they'd have transferred en masse to submarines that were not toying around with mines. As the Hellcat subs bore westward to the Tsushima Strait, the men's cockiness and profane outbursts began tapering off. Training exercises and emergency drills conducted at all hours of the day—flooding, fire, collision, chlorine gas—kept the men busy and primed to perform their duties with lightning speed and sure-handedness. There were no amateurs aboard submarines. The men of the Hellcats were for the most part veterans with war patrol experience who knew what to expect from one another and the Japanese—except for the part about the minefields in the straits.

Like all sub skippers, Earl Hydeman in the *Sea Dog* made it his business to tour his ship at least once a day to make certain that she and her crew would be ready for action when it came. In his Standing Orders Log he had posted the orders to be followed scrupulously by watch standers—especially the OOD—in all circumstances, as before long the ship would enter enemy waters.

In his standing orders Hydeman emphasized that officers of the deck "have the full responsibility for the ship and all the men in her. You should make the men on your watch feel, and act, their share of this responsibility also. As OOD, you are senior to all on board except the Commanding and Executive Officers; you are responsible, in turn, for keeping these officers fully informed."[1]

"I desire full reports of anything observed outside the ship [he wrote]; any untoward occurrence within the ship, or anything which changes the ship's military effectiveness, maneuverability in any manner, or ability to dive or surface. I do not desire reports concerning the ship's normal routine, unless for some reason it is not being maintained."

The Hellcats' normal routine got interrupted when the *Tinosa* of Risser's Bobcats made a detour to pick up the downed aircrew of a B-29, which they transferred to another submarine returning to Guam. On May 31, the *Spadefish* and *Sea Dog* received information provided by a U.S. patrol bomber and ComSubPac that another aircrew had ditched in their vicinity, east of the Nansei Shoto Islands. The two subs scoured the area but didn't find them. Hydeman ordered the pack to resume their northwesterly course to keep to their schedule.

During the northerly trek Hydeman made time to test the *Sea Dog*'s FMS on a pillenwerfer target. If he expected to see the decoy's cloud of exploding bubbles show up as a sharp green pear on the PPI scope, he was sorely disappointed. The laconic Hydeman reported, "Made dive for training and test of FM gear. Not satisfactory. Commenced repairs." Hydeman set the electronics gang to work on it, the OOD and watch standers high-stepping over parts and tools swaddled in clean shop rags on the diamond-patterned steel deck in the conning tower. Hydeman ran another submerged test early the next day. This time the retuned sonar gear demonstrated exceptional sensitivity, clear bell tones, and crisp, well-defined pears. But Hydeman's day turned sour when he learned that the SJ and ST radars had gone on the fritz.

In his patrol report Hydeman fumed over this development.[2] "This was a low for the life of the *Sea Dog*; having been plagued with many minor material problems* throughout the ship since the day after departure, we now lose our radar just before a scheduled transit through the Nansei Shoto." With two radar units out of commission, Hydeman faced a serious problem. Radar was absolutely essential to the operation; without its all-seeing eye the *Sea Dog* would be groping blindly as she approached the myriad of islands both big and small scattered throughout the Nansei Shoto, and later, as she made her run into the mouth of the Tsushima Strait.

Leaden clouds piling in threatened heavy weather when Hydeman rendezvoused with the *Crevalle* east of the Nansei Shoto to outline the radar problem he had. Steinmetz agreed to run interference for the *Sea Dog* as the two subs groped their way north through blinding rain.

Ten hours later Hydeman logged the following into his patrol report:

> Completed transit of the strait south of Akuseki [Island]. *Crevalle* did an admiral job of leading the blind, and is continuing to do so. We are communicating by VHF, following her signalled course changes, and managing to keep her wake in sight most of the time. Rain poured during most of the transit thru [sic] the strait. Fortunately, no [enemy ship] contacts were made by the *Crevalle*. Still working on the radars; have managed to get some results from the ST, but SJ refuses to revive yet. Informed *Crevalle* that completion of repairs by tomorrow was improbable, and asked his plans. He [Steinmetz] gave details and promised to look us up after surfacing tomorrow night and resume his duties as a "Seeing Eye dog" for us.

---

*The problems included a parted periscope hoist cable, a jittery gyro compass, a vibration in the auxiliary diesel engine, leaky lube oil coolers, noisy main motor commutators, and balky trim and drain pump motor controllers. The *Sea Dog* had undergone a refit at Guam prior to sailing. In his patrol report Hydeman pointedly remarked, "We pray for the day Sub Supply will get us working [replacement parts]."

As planned, the next night, June 2, the *Crevalle* found the *Sea Dog* and led her to a morning rendezvous with the *Spadefish* in preparation for their Tsushima Strait run-in. Just in time the *Sea Dog*'s radars came back online after a herculean effort by the radar officer and his technicians. At midnight the three subs crossed the imaginary line marking the southern approaches to the Tsushima Strait. Hydeman led his Hepcats into the western channel with the *Spadefish* in trail and the *Crevalle* off the *Sea Dog*'s port beam. Small craft of all descriptions buzzed across the strait like water bugs. Hydeman estimated that the Hepcats had a three-hour run through this traffic to reach their dive points. If they didn't run into Japanese patrols, and if their luck held and they didn't hit any mines, they'd be in the Sea of Japan by late afternoon of June fourth. Hydeman reminded his skippers by radio that their submerged transit would require nerves of steel and masterful ship handling. In regard to the latter, he cautioned that the inflowing Kuroshio Current possessed sufficient strength to push a sub sideways into the rows of mines dancing at the ends of their tethers. This was the real thing, not one of Lockwood's dummy minefields.

Hydeman kept an eye on the ship's clocks and the navigator's plot of their course into the straits. So far so good. A contact report from the *Spadefish* about a possible patrol boat didn't deter Hydeman; it was too late now to alter plans. Three hours past midnight on June fourth, the navigator tapped the chart lying on the plotting table. "We're there, Captain."

Hydeman had chosen a diving point assumed to lie many miles away from the first line of mines. It would give the three subs plenty of time to get into their positions in the strait before they had to run through the first line. Hydeman acknowledged the navigator's accurate calculation, pushed away from the chart table, and went topside. Weather conditions for the transit looked ideal: A low, easy swell under a light breeze wouldn't set the mines dancing like demons. Hydeman gave the order: "Officer of the deck, dive the ship."

"Dive the ship, aye." The OOD bawled into the IMC, "Dive! Dive!" then sounded the diving alarm.

The *Sea Dog* buried her bull nose in the night-black sea. On a northeasterly course, she began her run into the minefields that lay dead ahead.

Hydeman called away a quick test of the FMS that returned excellent results on a pillenwerfer tracked out to five hundred yards. Hydeman felt confident that if there were any mines up ahead, FMS would ferret them out.

"Make your depth one-one-zero feet," Hydeman ordered. "Maintain a two-degree up angle. All ahead slow." *This is it*, he could have added, but didn't, because every sailor aboard the *Sea Dog*, if not fingering a rosary or saying a silent prayer, was thinking the same thought their skipper was: *This is it!*

The Polecats and Bobcats weren't far behind the Hepcats. The six had transited the Nansei Shoto without incident and were closing in on their scheduled assaults on Tsushima.

During the long voyage, George Pierce in the *Tunny* had conducted training dives and mock attacks on Hepcats *Skate* and *Bonefish* to keep his crew sharp and alert for contact with the enemy. As the Hepcats stood toward Tsushima, they encountered heavy ship traffic, which called for broken-field running to stand clear. Many a juicy target had to be ignored. As Sieglaff had warned, an early attack would alert the Japanese that submarines had entered the area. Because the traffic was so heavy no one seemed to notice the three submarines in their midst, perhaps mistaking the Hepcats for small, fast steamers. The *Skate* and *Bonefish* rendezvoused with the *Tunny* to exchange information on the number and type of contacts they'd so far encountered, then set off on separate courses to avoid arousing suspicion, which three subs running at flank speed on the same course surely would.

Meanwhile, the *Flying Fish*, *Bowfin*, and *Tinosa* closed in on their objective from the south. The *Flying Fish* and *Bowfin*, too, had spent time searching for downed aircrews along the way, but none were found. Risser in the *Flying Fish*, eager to get to their destination and begin the transit into the Sea of Japan, had urged his pack mates on.

The Polecats and Bobcats, coming up behind Hydeman's Hepcats, timed their arrival so that Hydeman's group would have time to clear the strait, giving the remaining submarines a wide-open lane through which to make their transit.

Lawrence Lott Edge, commander of the
USS *Bonefish* (SS-223).

Sarah Simms Edge.

Lawrence Edge with his daughter,
Sarah "Boo" Edge, 1943.

Lawrence and Sarah Edge (*left*) with an unidentified couple at the Sir Francis Drake Hotel in San Francisco, California, January 1945.

18 Collier Road, Atlanta, Georgia.

USS *Bonefish* arriving in Pearl Harbor from a war patrol.

The *Bonefish* mascot
by A. L. Wilson, USN.

Edge receiving an award from ComSubPac chief of
staff Merrill Comstock.

The *Bonefish* crew. Lawrence Edge is kneeling in the center of the first row.

A Japanese patrol boat similar to those that sank the *Bonefish*.

Charles A. Lockwood, Commander,
Submarines United States Pacific Fleet.

Gaylord P. Harnwell, director of UCDWR and codeveloper of FM Sonar.

Chester W. Nimitz, Commander in Chief
United States Pacific Fleet.

Ernest J. King, Commander in Chief
United States Fleet and Chief
of Naval Operations.

Richard H. O'Kane and Dudley W. "Mush" Morton on the bridge of the USS *Wahoo* (SS-238). O'Kane served as Morton's exec before taking command of the USS *Tang* (SS-306).

Richard G. Voge, ComSubPac
operations officer.

Lockwood (*left*) greets Alexander "Alec" K. Tyree of the USS *Bowfin* (SS-287).

Postpatrol powwow: (*left to right*) Robert D. Risser (USS *Flying Fish* [SS-229]), an unidentified officer, Voge, Lockwood, and Comstock.

William J. Germershausen from the
USS *Spadefish* (SS-411) (*left*) and exec
Richard M. Wright with the
Spadefish battle flag.

Earl T. Hydeman of the USS *Sea Dog*
(SS-401), and task force commander of
Hydeman's Hellcats.

George E. Pierce, commander of the
USS *Tunny* (SS-282).

Everett H. Steinmetz, commander of the
USS *Crevalle* (SS-291).

Richard C. Latham, commander of the USS *Tinosa* (SS-283) with
Lockwood and Comstock.

Richard B. "Ozzie" Lynch
(USS *Skate* [SS-305]).

USS *Crevalle*, one of the nine Hellcat subs. Her look is typical of the
Gato- and *Balao*-classes.

A Japanese mini-sub on display at Camp Dealey submarine crew rest camp, Guam 1945.

UCDWR headquarters Building "X," San Diego, California, June 1944.

U.S. NAVY

The submarine base and tank farms at Pearl Harbor, Hawaii, circa 1944.

U.S. NAVY PHOTOGRAPH BY CHARLES FENNO JACOBS

U.S. Marines remove a blindfolded Japanese prisoner
from a submarine at Guam.

The heavy-gunned USS *Narwhal*, the first submarine Lawrence L. Edge served in.

William B. "Barney" Sieglaff.

A party prepares to board a junk (*seen in above photo*) typical of those encountered by U.S. submarines in the Pacific.

U.S. submarines alongside a tender, here the USS *Anthedon* (AS-24), in Subic Bay, Philippines.

The *Sea Dog* slowly felt her way into the western channel of the Tsushima Strait. The Kuroshio Current's inflow, pushing from astern, gave the sub a three-knot speed of advance. While this low speed of entry allowed for reasonable depth control, the current's push tended to crab the submarine sideways off course, as they had been repeatedly warned. The *Sea Dog*'s helmsman had his hands full steering the ship. Difficult as low-speed ship control was proving to be, should a mine suddenly loom up dead ahead, the submarine could be slowed and even stopped by reversing both props and backing down in time to avoid a fatal collision.

As the helmsman wrestled with the wheel, a sweat-soaked FMS sonarman scrutinized the sonar screen. Hydeman, watching over the man's shoulder, wanted to see bright green pears and hear hell's bells peal a warning that the sub had come into sonar range of the first mine string. He hoped to make contact sooner rather than later, to prove that the FMS was working. As the *Sea Dog* made steady progress through the strait, the glowing electronic contact beam swept around and around the PPI scope, its knife-sharp line so far uninterrupted by solid contacts. The deeper into the strait the *Sea Dog* penetrated, the quieter the ship became. Men exchanged inquiring looks: Was the damn thing working or wasn't it?

Hydeman had suggested, but not ordered, that men not on watch stay put in their bunks to reduce the confusion of having fidgety sailors with time on their hands moving about the ship. It would also conserve oxygen, as the submerged run would take at least sixteen hours.

The men were scared. It showed on their sweat-burnished faces. If the FMS wasn't working what chance did they have to survive their run through the minefields? None. A rumor shot through the boat: The FMS *wasn't* working; it never had worked, and the captain and officers had known it since leaving Guam.[3] Some men claimed they heard mine cables scraping down the sides of the *Sea Dog*'s hull. Others didn't hear a thing, only the soft hum of ventilation blowers. Some men feared that the cables that they thought they heard would snag on the sub and that her forward momentum would drag the mines down onto the ship, detonating them.

Two hours into the run the sound of distant explosions* rumbled through the sea. Depth charges? Or had the *Spadefish* and *Crevalle* hit mines? Then more explosions, this time heavy enough to shake the *Sea Dog*.[4] Was the *Sea Dog* destined to hit mines, too? Hydeman certainly didn't think so. Those undefined green blobs now blooming on the PPI—were they schools of fish or kelp? Could be either. It told Hydeman that FMS was working just fine and that there weren't any mines ahead of the ship, because the *Sea Dog* had lucked into a clear channel for use by Japanese shipping.

But sailors, even sub sailors, are superstitious, and to some of them on the *Sea Dog*, a sudden stroke of luck seemed too improbable. Gremlins—it had to be gremlins, the same ones that had attacked the ship's radars. Hydeman, the calm, steady-handed submariner, didn't believe in gremlins any more than he believed the ship's FMS would fail when needed most. Nevertheless, to provide a margin of safety in case the sonar unit suddenly went haywire, and recalling Sieglaff's stern warning to stay below two hundred feet in the event it did, he ran submerged as deep as he could, using the Fathometer to guide the *Sea Dog* over the strait's scarped and heaved bottom. These waters hid Japan's violent geologic history.

Sixteen tension-filled hours later, Hydeman made an announcement: The *Sea Dog* had entered the Sea of Japan. And to prove to all the doubters that the FMS was working just fine, he fired another pillenwerfer, which the PPI followed out to eight hundred yards. Hydeman raised the scope and described what he saw: a sea truck loaded with lumber, a large Japanese flag painted on her side, standing out from Korea toward Honshu.[5]

Hydeman surfaced the ship. Recharging batteries and air bottles, exchanging fresh air for foul, he set a course northeast for the *Sea Dog*'s patrol area off Honshu near the city of Niigata. The question the men

---

*The source of these explosions was never positively identified. One possibility was that the Japanese were dynamiting a site to install a shore battery either on nearby Iki Island overlooking the eastern channel, or on Tsushima Island itself. Another possible source was blasting in rock quarries on Tsushima Island.

aboard the *Sea Dog* were asking was, What had happened to the *Spadefish* and *Crevalle*? What about those explosions . . . ?

As it turned out the *Spadefish*'s transit, according to skipper Bill Germershausen, "was hair-raising but uneventful." The sub's FMS picked up mines, which the unit displayed with extraordinary clarity. Like the *Sea Dog*, the *Spadefish* ran deep, clearing the mines she encountered with ease. The real surprise came when the *Spadefish* surfaced after her sixteen-hour run and Germershausen saw Japanese ships steaming along the coast without escorts or air cover, red and green running lights ablaze as if it were peacetime. Germershausen could only watch them go by as he sprinted for his assigned operating area on Honshu's southwestern coast.

Everett Steinmetz had submerged the *Crevalle* in the strait, the *Sea Dog* off her starboard beam, the *Spadefish* astern. Steinmetz had had serious doubts all along about the efficacy of FMS, what with its inconsistency. Training on it at Pearl Harbor and Guam hadn't changed his mind. Now, however, Steinmetz was relying on it to locate mines showing up on the PPI in double rows at regular intervals. Okay so far, but, he grumbled, some of those mines hadn't registered until the *Crevalle* was within a hundred yards of them, too close for comfort.

As if double rows of mines weren't enough, the *Crevalle* passed under a pair of echo-ranging picket boats, their presence all the more sinister for a string of ship-rocking explosions, the same ones heard by the *Sea Dog* and *Spadefish*. Steinmetz said, "Although the [explosions] are not aimed at us they are close enough to cause the 'Lifted Eyebrow Department' to function every time."[6]

Despite mines and picket boats, the *Crevalle* made it on through. She surfaced surrounded by small craft, none of which seemed to notice that an American sub had, as if by magic, appeared in their midst. That they didn't was a mistake that they and their leaders in Tokyo would regret.

CHAPTER FOURTEEN

Threading the Needle

The *Skate*, a unit of George Pierce's Polecats, got off to a slow start from Guam. She'd barely cleared the antisubmarine nets guarding Apra Harbor when her skipper, Richard Lynch, received a salty report from the ship's engineering officer of a potentially disastrous breakdown. Lube oil in a reduction gear sump had dropped to zero through an open valve. After refilling the sump the motor macs saw the obvious: a port motor bearing had wiped out. The motor macs shifted power to the starboard shaft and replaced the bearing. In his report of the incident, Lynch wrote, "That very sound advice of that eminent Machinist's Mate, Simon McWrench should never be taken lightly: 'Don't never start no motors unless you got plenty orl in the berrins.'"[1]

The *Skate*, slowed but not stopped, caught up to the *Bonefish* and *Tunny* headed north. On June fourth, south of Tsushima, Lynch rendezvoused with the two Polecats to synchronize their submerged runs through the straits. Neither Edge in the *Bonefish* nor George Pierce in the *Tunny* had encountered any opposition as they approached the target.

The *Tunny* went through first. Her FMS worked perfectly, locating several strings of mines, which Pierce and his crew easily avoided. When he surfaced in the Sea of Japan he saw four fat, zigzagging targets. "What a temptation," he said, remembering that he had orders to hold fire until Mike Day.

The *Bonefish* followed the *Tunny*. She made a safe transit, but there's no record of how many mines she encountered and no report detailing the operational performance of her FMS. If Edge reported this information to Lynch or Pierce during one of two post-transit rendezvous, it isn't mentioned in their patrol reports.

The *Skate* followed the *Bonefish*, and, like the *Tunny*, encountered several strings of mines sown across the strait. There were so many overlapping strings that Lynch thought for a time that there weren't any gaps in the lines for his sub to get through. Then, one appeared. Squeezing through it, the *Skate's* bow brushed a mine cable. This was no phantom like the one heard aboard the *Sea Dog*; it was the real white-knuckle thing. Undaunted, Lynch pressed forward and somewhere between strings brought the sub to periscope depth for a look around and to try to get a navigation fix from a landmark ashore.

From his low-lying perspective, the scope just a few feet above the surface, Lynch observed a narrow and rocky coast overshadowed by sheer promontories. Massive and rugged, the western coast of Japan bore a striking resemblance to the coast of Maine. Its waters brimmed with small islands and narrow inlets. A network of lighthouses strung up and down the coast had in peacetime warned seafarers of the presence of dangerous rocks and shoals. Now, blacked out, their mute presence underscored the risks the Hellcats would have to run hunting targets along the coast. A submarine skipper audacious enough to enter these waters submerged would need skills bordering on the divine. Lynch dunked the scope and continued on, until the *Skate* stood clear of the strait, where it was safe to surface.

While the crews of the Hepcats and Polecats were breathing sighs of relief after their transits, Risser's Bobcats had yet to run the gauntlet. The *Flying Fish* went through first without a hitch. Early in the transit Risser became convinced that there weren't any mines—at least until the FMS started clanging and flashing after contacting solid mine balls on the sub's port and starboard beams. After evading this obstacle, the *Flying Fish* successfully dodged a patrol boat hanging around in the channel. All things considered, Risser thought it a rather routine operation.

Unlike the *Flying Fish*, the *Bowfin* encountered plenty of mines. Some of them had been planted less than sixty yards apart. Inching the *Bowfin* between them proved a tricky affair. In the conning tower Alec Tyree, eyes locked on the PPI scope, ears pricked for bell tones, searched for openings in the mine lines wide enough to conn the *Bowfin* through without snagging a cable. Tyree found them, but the stress it caused left him and his watch standers "mentally and emotionally exhausted by the experience."

Close calls by the *Bowfin* and the *Skate* proved simply a prelude to the *Tinosa*'s skirmish with what must have seemed to her skipper, Richard Latham, and his crew to be a deadly jungle of mine cables intent on trapping the submarine in an entangling embrace.

Ahead of the *Tinosa*'s penetration, Latham said he was feeling the effects that lack of sleep had brought on by long hours spent on the bridge. He agreed to a plan proposed by his exec, Lieutenant Commander Harvey J. Smith Jr., to spell each other during the most arduous part of the run. Several miles into the western channel, and while Latham slept, the *Tinosa*'s hell's bells started ringing just as the PPI's screen began throbbing with green pear-shaped blobs.

"Mines!" Smith and the FM sonarman huddled around the PPI screen, shoulders touching in the narrow space.[2] Green blobs had blossomed on either side of the *Tinosa*'s bow. As she approached the string, the blobs' shapes on the screen sharpened into deadly-looking pears. Smith knew from training at Guam that these were definitely mines, not schools of fish, patches of kelp, or anything else. Just big, solid mines.

Sailors ran sweaty palms over dungareed pant legs, licked dry lips. "Steady as she goes," Smith ordered. Every man knew that death danced unseen at the end of a cable just a few yards away, and that even a gentle brush against a mine's fragile horns would doom the *Tinosa*. But the *Tinosa* advanced and the mines slipped safely astern until at last the hideous green blobs faded from the screen and hell's bells stopped ringing. Even so the men at their stations heard blood pounding in their ears. No one spoke except for the duty telephone talker, who in a calm, steady voice gave the anxious men in other parts of the ship a running description of the action. "Mines are passing abeam. Signal's getting weaker by the second. Sonar's tracking the mines while Mr. Smith's conning the boat between them. Mines are astern now. . . . Everyone can relax."

But no sooner had they penetrated the first row than hell's bells started their frightful ringing again. The PPI scope displayed a seemingly impenetrable wall of green pears dead ahead. Smith voiced what they had learned during training exercises: Mines are never sown in solid rows.

Hearing a sailor's croaked, "There's always a first time," Smith counseled, "No, there's always a space between them. Keep looking for it."

The *Tinosa* inched forward. Smith felt the Kuroshio Current pushing hard from astern and knew that it would have the sub up against the mine wall in less than a minute. Too late to sheer away and start another approach, Smith was about to give the order to back down emergency when a gap appeared on the PPI scope.

With adroit manipulation of screws and rudder, Smith put the *Tinosa*'s bow into the gap between two mines. She squeezed on through, gingerly, like a person feeling for an open doorway in a dark room. In a few more minutes the mines would pass clear on either side. The men cheered and slapped one another on the back—

"Pipe down!" someone bellowed. "Listen."

Abaft the conning tower, the squeal of steel on steel began working down the *Tinosa*'s port side. It sounded like a sea monster clawing at the sub's hull. Mine cable! Men froze in their tracks. Would the welded-on clearing lines prevent disaster? If the cable snagged on something—a cleat, a diving plane, a propeller—the *Tinosa*'s forward motion might set the mine off or drag the mine ball down against the hull, breaking the dreaded horns, setting off the acid-filled detonators. Either way the *Tinosa* was a goner.

The young sailor coolly describing the action to his shipmates had lost his voice. Slack-jawed, he pivoted his head in the direction of the horrid, squealing sound like fingernails on a blackboard. His gaze had fixed, as had his mates', on the sub's curved pressure hull and its tangle of pipes and valves, where the mine cable seemed within arm's reach.

Smith faced a dilemma. If he tried to turn the *Tinosa* away from the cable she might get driven sideways by the current into the other mine cables hanging in her path up ahead. There was nothing Smith could do but wait for the squealing outside the crew's berthing compartment to work down the hull, where, a hundred feet aft, the submarine's stern diving planes and slowly turning propellers were waiting to snag the mine cable, to drag it down. The men trapped inside the *Tinosa*'s hull hardly dared to breathe as they prayed for the squealing to stop.

The cable crawled past the after engine room, where the hull tapered toward the stern and the diving planes and props. Time stood still. The cable was only inches away, little more than the thickness of the *Tinosa*'s hull, from sailors who expected at any moment to hear the explosion of a Japanese mine sending them to eternity. Instead they heard an eerie silence as the cable cleared the planes and props and disappeared into the *Tinosa*'s wake.

When Latham's head and shoulders appeared above the hatch coaming in the conning tower, it was all over; the sea monster had retreated. "Everything normal?" Latham asked Smith.

"Just a few mines, Captain. Now if you'll excuse me, I think I'll turn in."

When the *Tinosa* surfaced in the Sea of Japan, the submariners all agreed that they'd survived the worst of it and that from now on whatever they might encounter would pale by comparison. Young, resilient, in no time at all they had resumed their cocky, profane banter, even needling the cooks ladling out hot food. With the smoking lamp relighted and spirits high, someone broke out a worn acey-deucey board and a deck of greasy playing cards.

Latham sensed how eager his crew was, despite the restrictions on opening fire too early, to start sinking Japanese ships. They would, and soon. But before setting course for the *Tinosa*'s patrol station, Latham undertook some needed voyage repairs. With seas building, two men crawled forward onto the ship's plunging bow to free two mine cable clearing lines that had come adrift, fouling both bow planes. It was dangerous work that had to be carried out as cold seawater exploded over the main deck, soaking the huddled men fighting to untangle the thick, twisted cables that had carried away from their welded moorings. They also had to rig a homemade clearing cable in place of one that had been looped across the two forward deck cleats and which had also been carried away. Work completed, the exhausted, shivering men went below to hot showers, dry clothes, and warm food. Latham pulled the plug and took the *Tinosa* down to 120 feet to flee the now gale-force winds and give the crew a rest. Mike Day was two days hence. The *Tinosa* had a full load of torpedoes, and, like his men, Latham was eager to start shooting.

Meanwhile, in Guam the days since the Hellcats' departure hung heavy for Lockwood as he waited for a radio message from Hydeman about the status of the Hellcats.

Lockwood had outlined his plans for future attacks in the Sea of Japan in a letter dated June first and stamped TOP SECRET, which he sent to Frank Watkins in Washington. Before getting down to the business of his letter, Lockwood first thanked Watkins for the gift of six boxes of his favorite cigars, Crema Quintero. Lighting one up, he wrote:

> *Our first pack of 9 FM boats left here on 27–28–29 May, and we should be hearing from them on 9 June. I have absolute confidence in their ability to make it and my only regret is that we have not a second wave to put through about 1 July. I expect to meet them on their return at Pearl about 6 July and get the dope straight from the horses' mouth. . . . Recent developments and plans indicate that a very heavy concentration will be needed in the new hunting ground as soon as we can get it there. . . . The big thing to stress in placing pressure on the FM program is just that we need the maximum number which can be produced in the minimum time.[3]*

Watkins wrote back on June 9, 1945:

> *According to your letter you should hear from your FM Packs today. We have had our fingers crossed and a good grip on our left one, pulling for you. I too am confident that you will get good news.[4]*

Lockwood, savoring the fat Crema Quintero, blew a trumpet of smoke toward the ceiling in his office as he viewed the magnetic deployment board covered with submarine silhouettes. Yes, any minute now he might hear something from Hydeman. If the news was as good as Lockwood expected it would be, then he and Watkins—and everyone else, for that matter—could let go of their nuts.

PART THREE

Operation Barney

CHAPTER FIFTEEN
The Death of an Empire

June 9, 1945. Aboard the *Wakatama Maru* off the coast of Korea.

She'd been battling heavy seas all afternoon. As the weather slowly moderated, the navigator shot sun lines to get an accurate fix. The *Wakatama Maru* had an hour's steaming to make port at Bokuko Ko, Korea. For her crew it meant a good meal and, after loading the cargo awaiting her arrival, a decent night's sleep on a straw-filled mattress, not a hammock tossing like flotsam on a heaving sea.

Her crew had been having an easy time of it. Imperial merchant marine service, especially in the Sea of Japan, was easy duty. Though air raids by American carrier-based planes and long-range B-29s posed a constant threat, none of the ports in Japan where the *Wakatama Maru* off-loaded her cargoes of Korean lumber, ore, and rice had been hit. Best of all, the *Wakatama Maru* was a new ship, a proud ship, a good ship, not one of those old rust buckets pressed into emergency service for the emperor. Displacing 2,200 tons, she was over 300 feet long. Powered by modern triple-expansion engines and Parsons geared turbines, she could make turns for twelve knots.

The *Wakatama Maru* slowed as she approached Bokuko Ko. The sea bottom there rose abruptly inshore and a prudent seaman kept a weather eye on soundings for shoals. A moment's inattention and a ship could easily run aground—

Ba-whoomp! A tremendous explosion of flame, smoke, and pinwheeling debris erupted from the *Wakatama Maru*'s well deck. An exploding torpedo warhead lifted her out of the water, breaking her in two. Her amputated bow and stern halves hung suspended in midair for a brief moment before collapsing into the sea. Spewing sparks, smoke, and steam, the upended halves disappeared under a vast dome of swirling bubbles. A floating hatch cover, splintered planks, and a few struggling, oil-soaked men marked her grave.

Richard Latham had watched her come on, steaming blindly into range, blissfully unaware of the *Tinosa* lying in wait. "I must not let this guy get away," Latham thought.[1] And he didn't. He felt no remorse or pity for her crew; in submarine warfare there was room for only one sentiment: the satisfaction of the kill. The survivors were close to shore and would likely get there on their own before anyone realized that the ship was missing, as she'd gone down so fast there might not have been time to send an SOS.

The surprise attack phase of Operation Barney was now officially under way. But Latham had jumped the gun, violating Lockwood's orders to wait until sunset to open fire on Mike Day. Instead, he'd opened fire midday. Latham justified his action on the grounds that the other Hellcats were already in their assigned positions and that moving the deadline up a few hours and sinking a two-thousand-ton ship wouldn't give the Japanese an advantage they could use.*

To the east, across the Sea of Japan, Earl Hydeman in the *Sea Dog* inspected a harbor tucked between the two offset halves of Sado Island lying west of Honshu. Eager to sink ships, he nevertheless followed Lockwood's orders to the letter, waiting until sunset of Mike Day to open fire. Earlier in the day he'd held fire on a medium-size freighter that had passed within torpedo range. Hydeman didn't find any ships in the harbor, so he entered the narrow slot between Sado Island and the coast of Honshu. A high periscope trained east brought the coastal city of Niigata into focus.

---

*The Japanese at first dismissed reports of submarine attacks in the Sea of Japan. Then, even as it began to dawn that somehow U.S. subs had pierced the minefields ringing the sea, they were slow to grasp its significance and sound the alarm. By then it was too late: the Hellcats had infiltrated and the battle was on.

As if timed to Mike Day, while the sub was preparing to surface just after sundown the *Sea Dog*'s sound watch reported screws on a southwest heading. Hydeman turned the scope from its view of Niigata and saw a small freighter "running along serenely," as he put it, "on a steady course, side lights burning brightly."

"Make ready all tubes," Hydeman ordered. "Stand by forward."

The tracking party worked a quick, clean firing solution; the *Sea Dog* swung onto a ninety-degree track for a broadside shot.

"Match gyro forward, tube One!"

"Gyro twenty right, aye."

"Steady on one-one-zero!"

"Fire One!"

The *Sea Dog* lurched as the torpedo surged from its tube. Fifteen seconds into its run Hydeman motioned "up scope." Eight seconds later—*ba-whoomp!* He saw a flash of light and a thick spout of water at the ship's side. Heavily damaged, she went down in sixty seconds, leaving behind a solitary lifeboat loaded with men pulling oars, headed for shore.

There was barely time to savor the kill before another target, a freighter making nine knots, loomed up on radar. On the surface, Hydeman followed the ship against the glow of Niigata's lights. He saw the vessel's blocky silhouette looking like a flat black cardboard cutout pasted on the lighter-colored horizon. "Here he comes. Stand by forward . . ." And, "Fire One!" Ten seconds later, "Fire Two! . . . Fire Three!"

Three chalk lines—torpedo bubble trails—shot from the *Sea Dog*'s bow. The first torpedo missed and so did the second one, taking off on an erratic course of its own.[2] But Hydeman wasn't denied. A solid, satisfying hit from the third torpedo started a fire around the target's stern. Pandemonium broke out aboard the stricken ship. Panicked sailors ran to and fro while others calmly cranked out lifeboats from their davits. Though taking on water, the ship somehow managed to get under way again. "She's one tough bitch, ain't she, Cap'n?" said one of the sailors assisting the fire control party.

Hydeman agreed. "Fire Four!"

A hit forward of amidships slowed the freighter down, then stopped her dead. Her foremast toppled over, after which her bow dipped beneath the waves and broke off, leaving both midsection and stern angled skyward as they floated away in the dark.

Hydeman headed up the coast to find more targets before the Japanese realized that two ships were missing and sounded the alarm.

Mike Day was profitable for the Hellcats. Not only had the *Tinosa* and *Sea Dog* sunk ships, so had the *Crevalle*, working in an area north of where the *Sea Dog* had downed her two. The *Crevalle* had been prowling the coast as far north as the Tsugaru Strait, mindful of the mines dropped by B-29s. Like the *Tinosa*, she, too, had been plagued by mine-clearing wires coming adrift, which had fouled her bow planes, necessitating jury-rigging and, before dawn, the need to put men topside to effect repairs. Work completed, it was well after sunset on Mike Day when the call, "Radar contact!" alerted a crew already tensed for action.

Steinmetz horsed up the ladder into the conning tower—sure enough, there was a target on the radar screen: "Battle stations night radar!" He scrambled up to the bridge and saw an unescorted freighter with a distinctive raked bow lumbering into view against the land background. To Steinmetz, she looked as big as a house. Though the ship was blacked out, from time to time Steinmetz saw a bright light spill from an open door in her after deckhouse, which made her easy to follow.

Steinmetz swung the *Crevalle* around to bring her stern tubes to bear. Two torpedo hits flung flame and debris into the night sky. The *maru* staggered drunkenly, dropped her head, and went down under a pall of sooty black smoke. "Definitely sunk," Steinmetz reported.[3]

He sank another ship the next day, June tenth, an engines-aft cargo ship. He put two torpedoes into her, then watched her "turn turtle." In a sea of debris Steinmetz and his men on the *Crevalle*'s bridge counted about twenty-five survivors in the water clinging to a life raft. Earlier, Steinmetz had fired a single torpedo at an old-fashioned tug towing a raft of logs, only to see the tin fish skim along the surface like a playful dolphin and pass within inches of the tug's forefoot for a big miss. "Not so good," he said. This was better.

It was even better the next day, June 11—comical, too—when Steinmetz picked up another ship early in the morning.

> 0252 Target on northerly course . . . making about 7-8 kts. Manned Battle Stations. Target not seen from bridge until 5500 y. range. It's getting fairly light but we are both in a rain squall. Decided to try to get him on the surface. Headed in at 15 kts. . . . [F]ired two torpedoes to hit at M.O.T.

> 0312 Rang up all stop.

> 0310 Fired tube #5 and #6. Just a split second before firing, rain stopped and C.O. was about to order not to shoot. However, first fish was away on previous orders. . . .

> 0314 Target apparently sees us or the one torpedo that broached. Swung left with full rudder and went ahead flank. We went off in a cloud of smoke. Target blowing whistle and looks as if he is backing down. Smoke pouring from his stack.

> No timed hits. Started running ahead of target at 4500 yd. range. Realize that we should have gotten ahead and submerged in the first place. . . .

> 0325 Target in plain sight from bridge, binoculars not necessary.

> 0328 Submerged. Will try for a last chance gamble.

> 0331 At 58 ft. using ST [radar]. C.O. called angle on the bow 150 degrees port. ST called C.O. a liar as range is

closing. We couldn't believe our eyes. Nobody is that dumb. They must have seen us dive. Managed to calm down enough to get two tubes ready and swing to firing course. . . . This guy resumed his base course and kept on. He must have given the engineers hell as he is making ¾ of a kt more speed. His stack shows it.

0339 Fired tube #3. Fired tube #4.

0340-27 First torpedo hit.

-29 Second torpedo hit.

0342 Target started to turn over to port. There's a mad scramble to clear a life boat. Men are swinging out cargo boom on forward mast. Has about a 40 degree list.

0334 Target slowly righted itself but with considerably less freeboard than Lloyd's [of London] would insure. Bow section up to foremast looks like it might break off.

0344 These guys are game. They have opened up with about 5 machine guns from bow, bridge, and stern. All pointed in our direction. However, a periscope is a pretty tough target to hit.

0345 Bow is sinking slowly. When last seen, tracers were still coming from its machine guns. Stern is coming out of water. Target finally dove with 90 degree down angle. Reckon he will play hell pulling out of that. Considered sunk.

By now the Japanese knew that something extraordinary was happening in the Sea of Japan. At Pearl Harbor JICPOA had been waiting for an expected increase in enemy radio activity on or about Mike Day. It would signal that the intruding subs had been discovered and that the Japanese had gone after them. When that didn't happen, everyone breathed easier. Then, on the night of June 9, Pearl Harbor informed ComSubPac that it had intercepted an emergency transmission from a *maru* attacked by a submarine in the Sea of Japan. Lockwood lit a cigar. It had started. He had to hope the Hellcats would smash up everything in sight before the Japanese got themselves organized, and before all the *marus* fled open water for the safety of port. Lockwood wasn't overly concerned about Japanese antisubmarine measures. Fuel and manpower shortages, to say nothing of the lack of ships, would blunt efforts to go after the Hellcats. He was more concerned that targets would simply dry up, leaving the Hellcats wandering around, looking for something to shoot at. Impatient though he was, all he could do was wait for word from Hydeman that the Hellcats had started collecting for Mush Morton and the *Wahoo*.

Off the coast of Korea, the *Tinosa*'s periscope watch picked up the mast and top hamper of a ship heading for a port somewhere near Yangyang. Latham went after it, though it took more than five hours to work into position to attack.

The ship was a medium-size cargo ship with a red meatball flag flying from her flagstaff. "Fire . . . !"

Latham watched three white bubble tracks streak for the ship. "Down scope!" The setup was a perfect ninety-degree track, broadside to. Latham wagered that he had this one in the bag. "Up scope!" He looked for telltale flashes of light and geysers of water but didn't see a thing. All he heard was the sickening thud of a dud torpedo colliding with the ship's hull and, with that, the unmistakable sound of a torpedo's air flask exploding, not its warhead of deadly Torpex. Goddamn the torpedoes!

But there was something else out there, too. Something sinister. Something deadly: the rising pitch of high-speed propellers churning the sea. The sonarman, headphones clamped to his ears, jumped from his seat at the sonar console, terror written on his face. "A torpedo! One of our own! It's on a circular run! Back to us!"

The torpedo's horrid, whining scream resounded through the *Tinosa*'s hull. It was a submariner's nightmare come true. Something had gone wrong with the torpedo, maybe a jammed rudder or a broken gyro. Whatever the cause, the *Tinosa* had to get out of its way fast or risk being blown to bits. A circular-running torpedo was as dangerous as a child waving a loaded and cocked pistol in a crowded room, and just as unpredictable.

"All ahead flank! Take her down fast!" Latham ordered. The *Tinosa* clawed for the safety of deep water. As the sub plunged past two hundred feet, the men could hear but not see the runaway monster making ever larger concentric orbits until, at last, its chilling up-and-down Doppler scream slowly faded away. They exchanged nervous glances. Another close call, this time with a torpedo—one of their own, last time with a mine cable. The *Tinosa*'s hydraulic system moaned, as if protesting the unpredictable nature of submarine warfare. In the distance the boom, boom, boom of depth charges dropped by the freighter that Latham had targeted but not sunk seemed insignificant compared to the terrors of the circular run. Latham mopped his face on a towel, took a deep breath. "Okay. Let's see if we can find that guy who got away. Surface the boat!"

Despite a thorough search that took the *Tinosa* deep into the *Bowfin*'s area, Latham conceded defeat: The ship that had survived dud torpedoes and a circular run had vanished. As it turned out she stumbled across the *Bowfin* patrolling some seventy-five miles north of the *Tinosa*, off the Korean port city of Wonsan at Yonghung Bay. Alec Tyree in the *Bowfin* worked a setup on the errant ship and made the crucial run-in to the firing point.

"Shoot anytime, Captain."

Tyree did, four times. Then he turned his sub away and, looking aft past the *Bowfin*'s curling, boiling wake, saw a hit blossom on the target. Minutes later the stricken *maru*'s bow was pointing skyward. Armed depth charges left over from the string she had dropped near the *Tinosa* rolled off her slanting deck and exploded in the waters closing over her.

Snooping at night into Hokkaido's Ishikari Bay on June 8, the *Spadefish* paused outside the harbor at Otaru. Bill Germershausen considered the risks and rewards that might accompany a penetration of the harbor's mouth to get at the ships he saw in the roadstead seaward of the harbor's stone breakwater. There were more ships inside the harbor itself, but its confines made entry virtually impossible, to say nothing of the likelihood of grounding in shallow water.

Germershausen relied on radar and what he could see from the bridge through binoculars to locate any patrol boats hanging around the harbor entrance. The only vessel he saw was the station ship moored at the end of the breakwater. He didn't see any planes, even though the airfield at Sapporo had lights on for night operations. The lights cast an eerie greenish glow over the entire area, as did light from the city of Otaru itself. The stubborn refusal by the Japanese to follow simple blackout regulations demonstrated their lack of concern over air raids. Germershausen decided that the situation was stacked in his favor, that the rewards outweighed the risks of launching an attack at night in restricted waters.

Germershausen followed orders and waited until after sunset of June 9 to attack. As the sub headed in there was no need to call away battle stations other than the mere formality of it; everyone knew what the skipper had in mind and had been at their battle stations or hovering near them for hours.

The *Spadefish*'s raked bow cut toward the harbor and its moored targets. Radar swept ahead and seaward for ships and patrol boats, even as it outlined the harbor and, inland, the rugged hills south of Sapporo. As the *Spadefish* advanced she ran so close to shore that Germershausen smelled vegetation. He was grateful for a light rain that started falling, as it would mask the *Spadefish* slinking toward the harbor entrance by the breakwater. Meanwhile, planes flying into and out of Sapporo roared overhead. Their brilliant landing lights illuminated everything in their glide path, including the *Spadefish*. Germershausen and his men instinctively ducked for cover, though it was unlikely that the aircrews would realize that the sub they were overflying was American, not Japanese.

Halfway into the harbor Germershausen's plan fell apart. The ships in the roadstead turned out to be small ones. The only big one he saw was a freighter standing out to sea. She steamed right past the dark, low-lying *Spadefish*, on an opposite course, oblivious to her presence.

Germershausen had taken a huge risk and had come up empty-handed. The only target in sight on which to take out his frustration was the hapless station ship moored at the end of the breakwater. He fired two torpedoes, only to have them underrun the ship. Disgusted by their performance and by his overestimation of the station ship's draft, Germershausen issued a tangle of engine and rudder orders to spin the *Spadefish* on her heel and, at flank speed, haul after the ship standing out of Otaru.

Diesels roaring, Germershausen found the ship and charged in. Three crashing torpedoes ruptured her hull, blew her mast, stack, and cargo sky-high. She sank in less than a minute.

A half hour later radar picked up another ship, this one inbound for Otaru.

The *Spadefish* sped on, Germershausen tracking and plotting the target's blip on the radar screen. Three torpedoes holed this ship, too, which upended and, like the *Spadefish*'s earlier victim, went down in less than a minute.

And yet again, "Radar contact!" By now dawn was streaking the sky; it came early in these latitudes and Germershausen didn't want to be caught on the surface by planes dispatched to find the emperor's missing ships. He tracked this one by radar, then pulled the plug to finish it off submerged. Since there were no escorts to avoid, Germershausen kept the scope raised during the entire attack. It was almost too easy. All at once he saw a bright flash of light at the ship's waterline, then another, followed by a pair of trip-hammer-like explosions that rocked the *Spadefish* at periscope depth.

"Got her!" Germershausen turned the scope over for a quick look by the gang in the conning tower, who rarely had a chance to see a ship, broken in two, head for Davy Jones's locker.

George Pierce's Polecats—the *Tunny*, *Skate,* and *Bonefish*—arrived off Honshu's west coast on June eight. The *Tunny* began working an area around Wakasa Wan, due north of Kyoto, while the *Skate* and *Bonefish* started off around the Noto Peninsula, midway up the coast.

The mountainous Noto Peninsula formed a long hook that embraced Toyama Wan, a large bay bounded by small bays and bights suitable for use as anchorages. The tip of the Noto hook, known as Suzu Misaki, lies some fifty miles northeast of the interior cup of Toyama Wan. The bay, in some places, is over four thousand feet deep.

Ozzie Lynch in the *Skate* drew first blood for the Polecats on June tenth. Patrolling submerged off the Noto Peninsula, Lynch noted in his patrol report:

> Toyama Wan has two important seaports [Nanao and Fu-shiki] which are now heavily mined by our B-29s which will preclude our shooting anyone sitting in these places. Just north of [the Noto] peninsula is an island [Sado, in the *Sea Dog*'s op area] with an excellent anchorage. The important port of Niigata is just east of this island so that the Niigata traffic to Korea would be routed to pass just north of the Noto Peninsula. . . . [B-29 mining] has the effect of making traffic sporadic . . . and chases them into anchorages.

Unable to shoot at targets in those B-29-mined seaports, Lynch patrolled on the surface all night on the ninth looking for targets. In the morning he patrolled submerged along the fifty-fathom curve off the west coast of the Noto Peninsula. Around eight thirty in the morning Lynch spotted a small ship, which he avoided when it turned out to be a mine-sweeper. Then he saw something interesting.

Sighted a square black object on the horizon.

The damned thing has a gun. Battle stations submerged. It's a submarine. The sea is glassy calm.

1130 Fifteen or twenty degrees left zig. Right full rudder.

> **1144 Fired a salvo of four torpedoes. Looks like the
> I-121 [from ONI 208-J][4]**

Lynch's quick firing setup worked to perfection. Hit squarely amidships by two torpedoes, the big lumbering I-boat went down before smoke from the explosions had time to clear. A huge hissing bubble of air rising to the surface, followed by a glassy oil slick and breaking-up noises as she sank, signaled her end.

Lynch continued patrolling, avoiding areas that he knew had been mined by B-29s. On the eleventh he chased a large freighter, fired four torpedoes but missed. The racing freighter sought refuge in an anchorage filled with ships near Wajima. This anchorage presented a multitude of targets for Lynch, who went in after them.

"Stand by tubes forward!"

Six torpedo ready lights snapped on below the firing plungers.

"Up periscope!"

The exec, doing a two-step opposite Lynch at the scope, eyed the bearing marker around the scope's upper collar.

"Bearing, mark!" from Lynch.

"Zero-four-nine."

"Range, mark!"

The exec checked the data Lynch had dialed into the stadimeter repeater at the base of the scope. "Two-five double oh."

"Down scope!"

The tube started down. Before it reached the end of its travel, Lynch motioned "up." He snapped the handles down, steadied the scope on the targets. "Check bearing and shoot! Mark!"

"Zero-five-four."

"Fire One!"

The familiar blast of compressed air launched a torpedo at one of the Japanese ships riding at anchor in the harbor. At ten-second intervals five more whined from their tubes. Lynch reported that:

0912 Fired six torpedoes from the forward nest. The first five hit. . . . 1 hit in small AK [transport], 3 in medium AK, 1 in loaded medium AK.

0917 Largest AK sank. . . . Small AK settling, 1,500 tons or less.

0918 Gunfire. Pretty close. Fathometer shows 2 fathoms [twelve feet under the Skate's keel].

Japanese gunners had the Skate's periscope in their sights. Hot rounds raised water spouts all around it. Then:

0930 Here comes the opposition. A small vessel with a bone in its teeth. There's been a steady barrage of gunfire . . . and one came pretty close.[5]

More shells raised more water spouts around the periscope as Lynch sprinted for deep water. Depth charges dropped from the small charging vessel rattled the Skate but none was close. Lynch made a clean escape.

While the *Skate* was busy torpedoing ships, George Pierce, pack leader of the Polecats, had so far come up empty-handed. On the ninth, Mike Day, Pierce in the *Tunny* had hurriedly fired three torpedoes from long range at a steamer, only to be rewarded with a hopeless dud and two misses. Japanese sailors, hearing the dud strike the hull of their ship, Pierce reported, shined a light over the side to see what they'd hit.

On the tenth Pierce chased but lost a sub inside the B-29-mined reaches of the fifty-fathom curve where the *Tunny* couldn't go. Pierce could only hope that the *Skate* and *Bonefish* had gotten through the Tsushima Strait okay and that they were having better luck than he was.

———

Despite fighting heavy weather off the coast of Ch'ongjin, Korea, Robert Risser in the *Flying Fish* was so far having a better time of it than Pierce. His Bobcats had already sunk two ships and were busy tracking others. Risser was determined to run up the score as much as he could.

On June 10, the *Flying Fish* arrived at her patrol station at Seishin, an industrial port near the border between the USSR and Korea. Risser saw smokestacks, large buildings, and what looked like a refinery southwest of town. Fleets of fishing trawlers clogged the harbor entrance. As far as Risser could tell viewing the scene from more than a mile offshore, and given the jumble of godowns, sheds, trawlers, and whatnot lining the harbor, there were no large ships moored there.

Risser decided to stand off the beach to wait for something to show up. Around noontime his patience paid off. A sea truck, an eight-hundred-tonner, according to Risser,* stood out of the harbor under a sooty cloud of smoke belching from a single stack. Risser, still hoping for something better, tracked her until she was a good six or seven miles from the harbor.

> Fired two torpedoes. . . . First torpedo hit amidships . . . second missed astern. Target took a port list and settled slowly on an even keel. Crew manned boat and started to abandon ship but almost immediately climbed back on board. Target swung around to head for port and got up about 1–2 knots speed. They sighted scope and commenced firing at it with what appeared to be a 3" gun and a 20 mm machine gun.
>
> Fired a third torpedo to polish him off. I had no more than said "fire" when target took a sharp up angle and sank very quickly. Three torpedoes for 800 tons is not so hot! Two lifeboats with about 20 men in each pulled away for the beach.
>
> Surfaced [and stood] NE to parallel coast and cover [traffic lanes].[6]

———

*The eight-hundred-tonner turned out to be the 2,220-ton *Taga Maru*, a nice bag for Risser and a rare case of a sub skipper underestimating the size of his target.

Around midnight Risser heard, "Radar contact!" A small transport chuffing along on a zigzag course blew up when hit by a *Flying Fish* torpedo. Risser never saw her through the fog that had enveloped the coast until she lit up and exploded, her blip shattering into a thousand flickering stars.

Diesels burbling, the *Flying Fish* approached the target's grave.

> Heard shouting in water which stopped as we steamed slowly through wreckage. Secured from battle stations and slowed to steerage way.... [7]

It took two hours of searching through the wreckage and calling out to the survivors in Japanese before they coaxed a man aboard the sub.

> Took aboard one superior private [equivalent to a U.S. Army private first-class] of the Japanese Army.* He was the only one ... who responded to repeated calls of "Don't be afraid, climb aboard" in [Risser's] best Japanese. All others played dead on our near approach.
>
> Also recovered life ring and numerous charts. The charts turned out to be no value—all were old ones. This ship had done some cruising around Formosa, Hong Kong and Shanghai as evidenced by fixes on charts.†... Upon later questioning, our POW informed us that the ship was ... bound from Sakata [Japan] to Rashin [Korea]. She was a merchant ship, empty. Eleven troops were on board, apparently as an armed guard. Our man was a member of the 75mm gun crew of five

---

*According to Lockwood the prisoner's name was Siso Okuno. Because he believed that he had dishonored himself and his family, he wanted to commit hara-kiri. He was brought to Midway for interrogation and, like so many other Japanese POWs released after the war, simply disappeared.

†Exactly how these charts were recovered is not explained in the *Flying Fish*'s patrol report. In *Hellcats of the Sea*, Lockwood says "Risser's diving team" recovered them from the sunken ship's still-floating charthouse, this despite darkness and waters teeming with enemy survivors. The charts were supposedly dried out in the sub's engine rooms.

and the other six manned two machine guns. He can say "Thank you, sir" but professes no knowledge of English or [Korean]. Makes hen tracks beautifully and uses Arabic numerals. . . .

The hen tracks Risser referred to were those the prisoner had penned in a letter to Risser and the men of the *Flying Fish*. In it he expressed his shame at being taken prisoner and how deeply he longed for death, one that would exculpate his guilt for having lived. He admitted that he was surprised and heartened by the friendship and compassion shown him by the crew. Nevertheless, because he had been made to polish the sub's bronze torpedo tube inner doors and fittings, which he knew existed for the sole purpose of killing his people, he wanted to die all the more. He believed that his life was in the hands of America and that if he did die at the hands of Americans, it wouldn't matter, since he was already dead by virtue of his shameful deeds. He closed by thanking Risser and his officers and men for their acceptance of him as a man, and wished them health and happiness.

Here was the enemy made human, not just a gibbering, diabolical Japanese fanatic ready to plunge a knife into his enemy's heart. Submarine warfare, unlike ground warfare, inhibited direct contact with the enemy; most sub sailors rarely saw their victims up close. Robert Risser and his men could only wonder at the enemy they fought. Tenacious, fearless, deadly, madly suicidal on the battlefield, yet when taken alive and shown a dram of mercy, an enemy quick to acknowledge genuine gratitude, if not deep respect for their captors. As Lawrence Edge had learned from his encounters with Japanese prisoners, a Westerner could never hope to understand the Japanese and their warrior culture of Bushido. All an American fighting man could do was hope that a crushing defeat would eradicate forever the warrior mentality that had brought the Japanese and the world to the brink of disaster.

While Risser and the crew of the *Flying Fish* were interrogating the prisoner, the *Spadefish* was back on the hunt for targets, this time far to the north near La Pérouse Strait where her gun crews sank four small craft before Germershausen pressed on in search of bigger game.

Germershausen found it early on June thirteenth, fifty miles west of the southern tip of Karafuto, which, on a chart, has the distinct look of a giant crab claw. A distant solo radar contact, followed moments later by a second contact, pinpointed the position of two ships, both blacked out, one of them lying to, the other moving southwest. Were they Japanese or Russian? That the ships were unescorted meant nothing, for as the Hellcats had learned, few Japanese ships in the Sea of Japan had escorts. If these two were Russian, then according to the rules governing neutrality, they should be lit up with multicolored lights to identify them as such. Also, Russian ships crossing the Sea of Japan were supposed to follow a specific route that ran due west from La Pérouse Strait to Vladivostok. The ship contact heading southwest wasn't following this route. Germershausen peered from the *Spadefish*'s fog- and mist-swaddled bridge into an ink-black night, searching for lights moving or stationary, but didn't see a thing. He decided that the ship had to be Japanese. "Let's get him. Then we'll go back and get the other one, too."

The *Spadefish* closed in. Germershausen still couldn't see a thing. Steaming a straight course at five knots, the darkened ship took several minutes to draw ahead of the *Spadefish*. A shadow of doubt nagged Germershausen. He thought that if he could see the target instead of just its swelling blip on the radar screen, he could make sure that he wasn't shooting at a Russian. Then again, probably not; it was too dark and too misty. There were rules of the road even in wartime and they had to be followed. As Germershausen debated with himself—Japanese or Russian?—the TDC told him that he should fire torpedoes now.

A brilliant flash followed by a chest-thumping boom signaled the torpedoes' impact with the freighter. She circled, helplessly out of control, then, after stopping dead in the water, sank in just minutes.

Germershausen turned his attention to the second ship now suddenly lit up like a Christmas tree by lights identifying her as Russian. Germershausen suspected that the sight and sound of exploding torpedoes had frightened the Russian skipper into turning on identification lights to avoid a similar fate. Germershausen broke off the attack as the freighter got under way for La Pérouse Strait.

From a flurry of radio messages exchanged with Lockwood that began only hours after Germershausen had sunk the blacked-out ship, the chagrined skipper learned that he'd sunk a Russian ship, the eleven-thousand-ton former liner *Transbalt* of the maritime agency Sovtorg-flot. The Russians learned that she'd been sunk after a Japanese patrol boat responding to a distress call picked up her survivors. Domei, the Japanese government-run commercial broadcast network, announced that the *Transbalt* had been sunk by a U.S. submarine—deliberately, of course. The U.S. naval attaché in Moscow confirmed her sinking and this confirmation, as well as the Soviet government's angry condemnation, worked its way up to Admiral Nimitz. Nimitz and King didn't want the Japanese or Russians to know that U.S. subs were once again operating in the Sea of Japan, and so blamed the sinking on a Japanese submarine. The Russians weren't fooled, but given their need for U.S. military aid, they dropped their protests and everyone breathed easier, especially Lockwood.[8]

By June 14, Earl Hydeman had received radio reports from his Hellcat commanders describing their success (or lack thereof) sinking ships, including their torpedo expenditure thus far (the nine Hellcat subs carried a total of 216 torpedoes). Hydeman also had a fairly clear picture of what the Japanese were doing to thwart attacks by the Hellcats, and it wasn't much. To keep the raiding subs at bay, the Japanese were rerouting ship traffic inside the fifty-fathom curve and even the twenty-five-fathom curve. Hydeman immediately altered the packs' assignments, pulling them back from deep water, where ship contacts had dropped to almost zero, and instead deploying them into safe coastal areas where ships were taking refuge. The information he received also pointed to problems that had so far reduced the Hellcats' overall effectiveness, mainly torpedo problems. Adding to it were the rushed, failed long-distance torpedo shots fired by overeager skippers, and targets that seemed to magically elude destruction by a sudden and fortuitous zig or zag. Hydeman may have been disappointed by these mishaps, but he surely wasn't disappointed that the Japanese countermeasures they had all anticipated hadn't yet materialized. So far attacks on the Hellcats had been perfunctory at best. To be sure, no one was underestimating the Japanese; there was plenty of time between Mike Day and Sonar Day for them to launch an all-out assault on the Hellcats, including a blockade of the strait. Yet, if what the Hellcats had experienced so far was any clue to what the Japanese had in mind for them, there was little to worry about. If Hydeman and his skippers had had the time and inclination to ponder this they might have concluded that the inability to quickly mount a strong response portended the imminent defeat of the empire.

CHAPTER SIXTEEN
A Dark Silence

Lawrence Edge had conned the *Bonefish* past the last string of mines in the Tsushima Strait and entered the Sea of Japan late in the day on June 5. His patrol area coincided with those of the *Tunny* and *Skate*, around the Noto Peninsula, but Edge didn't make contact with his pack mates until the fifteenth, when Ozzie Lynch in the *Skate* raised the *Bonefish* on SJ radar sometime around midnight.

Rendezvousing, Edge reported that he'd been busy. On the thirteenth he had sunk a large cargo ship. This bag would turn out to be the biggest of the Hellcat foray. The *Bonefish*'s torpedoes slammed into her as she was steaming north of the Noto Peninsula, not far from the area in which Hydeman's Hepcats had been working with such success. So far the Polecats were in the lead: In eight days the *Bonefish* and *Skate* had sunk four surface ships and a sub.

Late on the fifteenth, after George Pierce exchanged recognition signals with the *Skate*, he radioed both her and the *Bonefish* to rendezvous for their patrol assignments after dawn on the sixteenth. The *Tunny* arrived as scheduled and Pierce saw the *Bonefish* on the surface, waiting, but not the *Skate*. He assumed that the *Skate* had gone off somewhere on her own and had missed the rendezvous. Radio communications were spotty, what with the Japanese trying with limited success to jam the Hellcat frequencies. Sometimes it was more efficient for the subs to lay to alongside one another and exchange news by shouting through megaphones.

Pierce, yet to sink a ship, was pleased with Edge's successful torpedo action. He gave Edge a new patrol assignment in an area around Suzu Misaki, at the tip of the Noto Peninsula. Rather than wait for the *Skate* to show up for her assignment, Pierce radioed instructions to Lynch. He received Lynch's acknowledgment, after which he shaped a course to patrol the traffic lanes in and out of Toyama Wan.

On the sixteenth the *Skate* and *Bonefish* rendezvoused on their own to exchange more information. Lynch reported that he had picked up and tracked three patrolling escort-type ships on radar. Out of torpedoes, he gave the trio a wide berth. Edge confirmed that he, too, had had brief radar contact with the same escorts. He thought that they were searching for the sub—the *Bonefish*—that had sunk that big *maru*. Lynch told Edge that without torpedoes all he could do was keep an eye open for targets and try to vector Pierce and Edge into position to attack them. If not, he would head north to await Sonar Day, as the Hellcat op orders had instructed. Edge left the decision up to Lynch, gave him a "roger," and shoved off to patrol his assigned area around the Noto Peninsula.

Farther north, the unflappable Earl Hydeman had been busy dealing with the gremlins that had plagued his ship from the start. This time a mine-clearing cable had wrapped itself around the *Sea Dog*'s starboard propeller shaft. It had to be fixed, as the loud thumping noise and heavy vibrations coming from the shaft might lead the Japanese right to her. The question was, What could be done at night in icy water and rough seas to clear the one-and-a-quarter-inch steel cable fouling the shaft?

While the *Sea Dog* lay to, two men, one of them the chief of the boat, went over the side into numbingly cold water, equipped only with crude diving gear and heavy-duty bolt cutters. It was dangerous working topside at night. The hissing seas engulfed the men, who shivered and clung to hull protrusions with fingers stiffened by the cold. Worse yet, their diving masks leaked so much that they couldn't work underwater. Finally, the bolt cutters proved useless on such thick cable. Forced to admit failure, Hydeman ordered the two men back aboard and resumed his course. He was convinced that the noise and vibration coming from the shaft, which slowly diminished over time, was, like the ship's SJ radar, cursed by gremlins, if not another low point in the life of the *Sea Dog*.

Undaunted, Hydeman continued tracking and torpedoing ships.

Pursuing three ships steaming along the coast, he fired and hit the leader, only to have the other two make a run for it. Before Hydeman could give chase, a plane appeared. Thinking he had plenty of water under the keel, he took the *Sea Dog* down fast and felt the jolt of a solid grounding in shallow water. Gremlins had struck again.

"Flooding in the forward torpedo room, Captain!"

The grounding had wiped off the ship's two bottom-mounted QC soundheads; water poured into the torpedo room through the wrecked soundheads' gate valves.

"Take charge below," Hydeman ordered his exec, who laid forward on the double to assess damage and lend a hand.[1] Then to the diving officer, Hydeman said, "All back full; shift the rudder. We'll back her off."

They did and left behind both soundheads neatly sheared off at the shaft flanges. Somehow the pilot of that plane hadn't seen the stranded *Sea Dog*'s dark outline in shallow water. If he had, it might have spelled her end.

Aboard the *Crevalle*, Steinmetz, after sinking three ships on successive days, hunted for more targets. On the thirteenth he ran across a pair of luggers, both loaded to the gunwales with cargo. They were too small for torpedoes, but perfect targets for the *Crevalle*'s guns.

Steinmetz called away battle surface. The *Crevalle*'s gunners rushed topside, unlimbered the sub's five-inch and 40mm guns, and began slamming rounds into the hapless luggers, smashing them to kindling, driving their crews overboard. Their hulls shattered, strakes exposed like the ribs of dead animals, cargoes scattered over the sea, the two luggers staggered under the onslaught, capsized, and sank. Shooting motion pictures from the bridge, the *Crevalle*'s photographer's mate captured the action from start to finish on color film. Not content with what they had, Steinmetz moved in for close-ups. Lockwood wanted footage of what his submarines were doing, and Steinmetz would give it to him. The *Crevalle*'s skipper noted in his patrol report that

[We] thought we detected a Nip hiding under a box floating near the target, so heaved a couple of hand grenades at the target and took a few pot shots with carbine and sub machine-gun at box.

Destroying enemy vessels was one thing, but trying to kill an un-armed survivor in the water was an altogether different matter, one not sanctioned by ComSubPac for obvious reasons.[2] Steinmetz, perhaps real-izing this, called a cease-fire and moved on, looking for bigger game.

He found it on the fourteenth. North of the Tsugaru Strait, Steinmetz picked up three merchant ships hugging the coast and guarded by two pinging destroyers. Working into position between the *marus* and the destroyers, Steinmetz fired at one of the *marus* and missed, which brought a nasty two-hour depth charging. Undamaged and running silent, the *Crevalle* crept away, into deeper water.

Across the Sea of Japan, Alec Tyree in the *Bowfin* had sunk his second ship, an eight-hundred-ton engines-aft freighter. She went down, leaving a lone survivor clinging to a capsized lifeboat. Tyree continued searching for, as he put it, greener pastures along the fogbound Korean coast, but found only fishing boats. On the fourteenth the *Bowfin*'s gunner's mates shot up a twenty-ton two-masted schooner. Four days later Tyree ran smack into a pair of patrol boats hidden by fog. Despite severely reduced visibility, the boats landed shells just yards away from the fleeing *Bowfin*, their fragments sleeting across her bow. Tyree pulled the plug and gave them the slip.

Things were no better farther north along the east-west shipping lanes to northern Honshu. Hampered by swirling fog seemingly too opaque even for seagulls to fly in, Tyree had to patrol for targets through fleets of fishing boats with their entangling gill nets. He wanted another crack at a decent-size ship before Operation Barney came to an end. But, now that the Japanese had warned merchant ship captains about the pres-ence of U.S. submarines in the Sea of Japan, cross-sea shipping had dried up. Instead of cargo ships, Tyree had run into more and more radar-equipped patrol boats.

On the twentieth, Tyree encountered a large southbound engines-aft freighter with two escorts in trail. He worked in on the surface and fired six torpedoes. Sound tracked them to the target, but all Tyree heard were loud explosions from what sounded like depth charges and gunfire, not torpedo warheads. Cursed by the never-ending torpedo nightmare, he tallied six misses, one of them a circular run. "Sighted all targets going away with no apparent sign of damage," he said. "A sad sight."[3]

Tyree refused to give up. He started a flank-speed end around that took the *Bowfin* through a large fleet of sampans, which, Tyree discovered in the nick of time, masked one of the freighter's escorts laying back to find the submarine that had fired those torpedoes. The escort took out after the sprinting *Bowfin* and stayed on her tail until Tyree submerged and went deep. Free of the escort, he doubled back to search for the freighter, which, like so much else about coastal Korea, had disappeared into a gray, swirling oblivion.

Richard Latham in the *Tinosa* had had no better luck than did Tyree at this stage of Operation Barney. Though he failed to sink the ship that had fled into the *Bowfin*'s area, where she was nailed by Tyree, Latham continued to hunt for ships worthy of a torpedo.

Less than a mile off the Korean coast he encountered a large sea truck, which the *Tinosa*'s gunners quickly demolished, including its crew, who frantically tried to load and fire a gun from the sea truck's bow. After that, targets dried up. Like the *Bowfin*, the *Tinosa* mingled with fishing boats and spit-kits of varying sizes, groping through dense fog and ice-cold rain, hunting targets that had seemingly vanished.

Then, on June 20, the *Tinosa* was inching north for the Hellcat rendezvous on the twenty-fourth when she picked up a small freighter, which she blew to bits with three well-placed torpedoes. The target never slowed, just kept right on plowing down, down, down until she disappeared in a cloud of steam from her exploding boilers. Latham spotted a lone, shocked survivor clinging to wreckage; the ship sank in less than a minute.

At sunset Latham hit a tanker loaded with aviation fuel. An enormous, rolling fireball engulfed the ship, lighting up the sky like a setting Japanese sun. Her sinking, Latham's fourth, capped a successful patrol.

Earlier, the *Flying Fish* had been faring no better than the *Bowfin* and *Tinosa*, searching for targets among the rocks and fog of the coast of Korea. Like Tyree and Latham, Risser had doubts that the area would produce anything worthwhile. He had so far sunk only two ships, though he had fired at others and missed, due mainly to unfavorable attack positions. He was still aiming to run up his score. At dawn on the fifteenth, outside Seishin Harbor near Ch'ongjin, Risser tore into a fleet of sampans loaded with bricks destined for a construction project around the harbor breakwater. The *Flying Fish*'s guns wrecked and sank ten sampans and damaged two tugs towing barges loaded with boulders.

Risser, astonished that the ruckus didn't bring out patrols from inside the harbor, chanced a peek through the periscope past the breakwater. He saw ships at anchor in the shallow inner harbor, but he had no way to get at them. Later, Risser missed a shot at a transport headed downcoast for Seishin. After the transport hightailed into the harbor, a Japanese escort vessel ventured out and began dropping random depth charges. By then the *Flying Fish*, back on the surface, was on her way to join the *Bowfin* and *Tinosa* ranging northeast toward La Pérouse in preparation for their pre-exit rendezvous.

After the *Spadefish*'s attack on the *Transbalt*, Germershausen finally found a legitimate target. Just north of La Pérouse Strait, he sank a two-thousand-ton *maru*. Moving south to Tsugaru Strait, the *Spadefish* then torpedoed another two-thousand-tonner. After that, things settled down. Germershausen had no reason to complain: He'd sunk five ships (six if one included the hapless *Transbalt*) and four small craft. He was content to wait for Sonar Day, when the Hellcats could break for home. On the other hand, if something came along while he was waiting, he'd try to run up his score. Germershausen didn't let his guard down. The Hellcats couldn't allow themselves to fall victim to complacency or hubris. That would be a fatal mistake. While the Japanese defenses had stiffened, they hadn't yet mounted a determined effort to find and attack the Hellcats. That could change.

To the south, the big ship sunk by the *Bonefish* had spread a carpet of debris and survivors over a huge area north and west of the Noto Peninsula. Currents and tides had pushed the flotsam inshore, where George Pierce, patrolling the traffic lanes off Toyama Wan, encountered more than a dozen life rafts loaded with oil-soaked and shivering survivors. As the *Tunny* approached a raft, the survivors lay facedown, apparently expecting to be machine-gunned. Despite Pierce's exhortations lifted from a Japanese phrase book, none of the survivors could be coaxed aboard. Pierce gave up and moved on.

Early the next day, after combing the area for targets, Pierce spotted the same drifting rafts still clotted with survivors. He surfaced in their midst, no doubt a frightening scene for the exhausted onlookers—the sea parting and hissing, a submarine rising from the depths shedding white water. This time Pierce took aboard a senior Japanese navy petty officer who'd apparently had enough raft time and was eager to cooperate with his rescuers in exchange for dry clothing and hot food.

Resuming the patrol, Pierce received a radio message from Lawrence Edge, stating that he had the masts and stacks of two ships in sight and giving their coordinates. Pierce bent on four mains to find them and join the *Bonefish* in a coordinated attack.

It was after full dark when the *Tunny* and *Bonefish* made radar contact with the targets. After Pierce and Edge planned their attack via voice radio, the two subs closed in. Edge reported that he was within a thousand yards of the targets but still couldn't see them. Darkened ships and surface haze always made a night surface attack difficult to execute. Yet there was something else out there, something spectral and sinister evidenced by a wobbling electronic shimmer on the SJ radarscopes of both subs. Shades of Harry Greer and the *Seahorse*: Japanese radar interference!

Pierce and Edge smelled a trap. The two ships they had been targeting, which up until now had seemed unaware of the presence of the two subs, suddenly opened up with their guns. Patrol boats!

Four-inch rounds started dropping around the *Tunny*. As Pierce and Edge veered away and hauled out at flank speed, the two patrol boats began dropping random depth charges. Both subs went deep to let things cool off. They'd almost been suckered by a hunter-killer team. The next morning, June 18, Pierce, relieved to see that the *Bonefish* had escaped from the patrol boats, held a confab with Edge via megaphone. After reviewing the trap set by the patrol boats,

> Larry asked for permission to make a daylight submerged patrol in Toyama Wan. Decided to split up for independent operations close to coast tomorrow. Set course for Wakasa Wan; Bonefish set course for Suzu Misaki.[4]

Edge's asking Pierce for permission to patrol in Toyama Wan was standard procedure, as it was his duty to keep Pierce informed of his movements at all times in order that the actions of the *Tunny* and *Skate* could be properly coordinated with those of the *Bonefish*. Edge was merely following regulations, which required a junior officer (Edge) to obtain authority from his superior officer (Pierce) to search his assigned area, a task no more nor less risky than any of the other operations, including harbor penetrations, that had been conducted by the Hellcats. In any event, the forty-mile-long-by-twenty-mile-wide Toyama Wan had the potential to contain targets that an aggressive skipper like Edge was not about to ignore.

After her rendezvous with the *Tunny*, a mysterious, dark silence enveloped the *Bonefish*.

For a time the *Bonefish*'s disappearance went unnoticed by Pierce and Lynch. On the eighteenth and nineteenth, Lynch, target spotting for his mates, reported seeing two small transports, an ancient coal-burning destroyer, and a small R-class sub, near the mouth of Toyama Wan. When his initial contact report went unacknowledged, Lynch tried to raise the *Tunny* and *Bonefish* on radio. Neither of them answered, and Lynch concluded that they were patrolling submerged. When Pierce finally showed up the next day, he radioed Lynch to start heading north to La Pérouse for the prearranged rendezvous with the other Hellcats. In their exchange Lynch didn't report hearing any explosions from torpedoes or depth charges.[5] Pierce also radioed Edge with similar instructions, but didn't receive an acknowledgment. Edge's silence on the radio net didn't raise an alarm with Pierce because he assumed that Edge was busy hunting for targets in Toyama Wan and that he'd catch up with the other Polecats later.

Heading north, Pierce, who had yet to sink a single ship, ran down a large escorted freighter he estimated was in the ten-thousand-ton range. Out of position for a close-in attack, he fired a long-range shot and missed. Pierce was the only Hellcat skipper who hadn't sunk a ship, and if he had any hope of making up for it, time was fast running out.

It was now June 21. The Hellcats had staged a successful raid into the Japanese bastion, though the final tally was yet to be written. They weren't home free, but so far had met and overcome every challenge laid before them. They had also demonstrated conclusively how versatile submarines were and how valuable that versatility would be in the future, when subs would be called upon to undertake missions far more dangerous than Operation Barney. Not one to show his feelings, inwardly Hydeman was pleased and impressed: His Hellcats had performed magnificently. He radioed them to head for La Pérouse for the breakout.

CHAPTER SEVENTEEN
Breakout

espite newspaper headlines announcing imminent victory in the Pacific—"Jap Defeat Assured"—Sarah Edge wasn't fooled into thinking the war was over for Lawrence and his men. Indeed, she waited every day for a letter saying that he had returned from patrol, that he was safe and that his arrival meant he was destined at last for shore duty.

Lawrence received a letter from Sarah before sailing from Guam with the Polecats on May 27. She played a little guessing game based on a popular movie* she'd seen in an Atlanta theater about submarines.

> *Dearest Shug[ar],*
>
> *I can hardly wait to get home to get a letter to see just what you will be doing. If I do not get one today then I am sure of it tomorrow. Wish I could guess the type of mission you were on. Of course, it could be 1) catching our aviators, or 2) it could be something like "Destination Tokyo"—going through an [anti]sub net into a Jap harbor to let spies ashore and then waiting 'til they*

---

*A rough draft of this letter is the only letter from Sarah to Lawrence that exists. None of the hundreds of letters she wrote, which Lawrence kept, nor any of his personal effects, were returned after his death. Each submarine had a storage locker assigned to it aboard the submarine tenders, but the *Bonefish*'s locker aboard the USS *Apollo* (AS-25) at Guam was reportedly empty. Lawrence most likely kept Sarah's letters aboard ship.

The movie was *Destination Tokyo*, starring Cary Grant, John Garfield, and Dane Clark. In the film an American sub penetrates Tokyo Bay for the purpose of putting agents ashore to collect weather information.

had their data and bringing them back home again, or 3) it
could [be] something else of which I have no idea. Wish I knew.

When Sarah received the letter Lawrence had posted before his departure from Guam on Operation Barney, he was far out at sea.

May 25, 1945

Most darling wife,

*This, I fear, because of the press of time will just be [a] short
letter [to tell you] that I love you so deeply and completely. . . .*
*. . . There's just a spot of news this morning. . . . We had a
surprise in that word was received that our last patrol was
called "successful" because of the special mission, although we
had not guessed it would be so. Anyhow, it is a pleasant surprise
as at least the Bonefish string of successful patrols had not been
broken after all, and our new men and officers can have
[combat] pins after all.*
*Also unexpectedly [an officer] developed a kidney stone two
days ago and the doctors have transferred him; so now we have
two new officers to replace him and [another officer]. The new
ones have just reported aboard, so I can't say that I know them
yet. . . .*
*. . . Darling, I guess it's because I wished so hard . . . another
letter from you came this morning, plus one from Mother. None
had come for about three days and I was sorely looking for one
more. Mother wrote about you and Boo's bringing them my
picture on Mother's Day, and how much she and Dad both
appreciated it.*
*Goodbye, my precious for today. You'll be constantly in my
thoughts as well as my heart until I can write again—and for
always.*
With all my deepest love,

Lawrence

On June 18, 1945, President Truman convened a meeting in Washington with the Joint Chiefs of Staff to discuss an invasion of Japan, should it become necessary. Those charged with its planning believed that it was imminent. Truman was deeply troubled by the shockingly high casualties American forces had suffered at Iwo Jima and at Okinawa. The number of dead and wounded in those two campaigns alone portended a grave outcome for any assault on Japan itself.* The casualty estimates varied wildly, depending on who was preparing the estimates; the prospect of hundreds of thousands of U.S. dead and wounded in an invasion shocked Truman. In the event, the casualty issue wasn't settled to his satisfaction. All he and the Joint Chiefs knew for sure was that it would be frightfully high. Whether or not it would be too high and therefore unacceptable to the American people was left unanswered.

Truman had only two choices when it came to ending the war. One, he could order an invasion and then prepare the American public for the carnage that would result. He and his chiefs never doubted that the United States would prevail in an assault on the home islands, but they knew that an attack on the core of Japan's power and culture would provoke a response from its army and civilian population far more bloody than any so far encountered in the entire Pacific campaign. Truman's second choice was to authorize the use of the atomic bomb on Japan. The bomb would be ready for testing at Alamogordo, New Mexico, by mid-July, though Truman had been cautioned that there was no guarantee it would work. If it didn't, he'd be faced with ordering an invasion.

A third alternative, albeit an unpalatable one, was to offer Japan a negotiated peace in lieu of unconditional surrender. This alternative had been suggested by more than one respected public figure. Such an offer was wholly unacceptable to Truman and his advisers and to America's

---

*For the invasion of Manchuria the Soviets made provisions for dealing with upward of 540,000 casualties, including 160,000 killed (a mere drop in the bucket compared to the millions of Russians slaughtered fighting on the Eastern Front). The numbers were based on Soviet assumptions of Japan's intention to fight to the death, the same assumption that the United States had made from its appalling losses at Iwo Jima and Okinawa.

allies. The only way the Japanese would surrender, they believed, was for the United States to utterly defeat them. This, of course, was Lockwood's view of Operation Barney: Smash the Japs and end the war *now*.

While the president and his advisers wrestled with these issues, work on the atomic bomb progressed at a steady pace. As the Hellcats were invading the Sea of Japan and sinking ships, scientists at Los Alamos, New Mexico, began assembling the parts necessary for a uranium-fueled atomic bomb. Meanwhile, radioactive plutonium produced in the nuclear reactors at Hanford, Washington, and machined into a two-part hemisphere smaller than an orange, arrived in Los Alamos for use in a bomb scheduled for testing on July 16.

Back in February, a young naval weapons specialist had been dispatched from Admiral King's office to brief Admiral Nimitz on this new weapon. King had received his briefing directly from the White House and had instructions to pass the information to Nimitz and MacArthur. In his explanatory letter to Nimitz carried to Guam by the weapons specialist, King gave CinCPac a highly simplified explanation of the bomb and its potential destructive power. He added that it would be available sometime in August. Nimitz, after reading King's letter, listened politely to the weapons officer's technical explanation of how the bomb worked, thanked him, and got on with his job. To Nimitz, faced with the hard realities of fighting a war with conventional weaponry, talk of an atomic bomb seemed like just so much science fiction, if not the stuff of pure fantasy.

For his part, Lockwood had no knowledge whatsoever of the atomic bomb. Its secret was closely held among only a handful of top military commanders, like King and Nimitz, and civilians, like John J. McCloy, Truman's assistant secretary of war and a close personal adviser. Even if Lockwood had been told about the bomb, it's unlikely he'd have thought that its immense power used on the Japanese would end the war and end the need for any more Operation Barneys. Even before he'd heard the results of Operation Barney, and before the Hellcats' return from the Sea of Japan, Lockwood was pushing hard for more FMS-equipped boats to send in there.

In a memorandum to his staff dated June 23, Lockwood wrote that he expected to find that some of the Hellcat skippers went very deep through the Tsushima Strait minefields, essentially disregarding their FM sonar. He expected that this would inspire a lot of "dare devils to go and do likewise with no FM at all." Finally, he wrote, if it appeared safe to do that, he'd entertain such offers during the period when the supply of FMS boats was low. "I want to get more boats in there in July if humanly possible. Therefore please circulate the word as to volunteers. . . ."[1]

Lockwood had been following the Hellcats' progress via FRUPAC's (Fleet Radio Unit Pacific) intercepts of Japanese radio messages and from bits of intelligence pieced together by JICPOA. He'd even pored over photographs taken by the Army Air Force during reconnaissance missions over the Sea of Japan, which had returned images of the three ships sunk by the *Skate* sitting on their bottoms in the littoral waters of a harbor at Noto Hanto.

From these sources Lockwood began to sense that so far Operation Barney had been a huge success. In particular the Japanese, besides broadcasting alerts to antisubmarine units and merchant ships in the Sea of Japan, had let slip over Domei that U.S. submarines had attacked ships in the sea and that they in turn had been tracked down and sunk off the western coast of Honshu. The reports didn't say how many subs had been sunk, but the implication was that there were a lot of them—many more than just nine Hellcats.

Lockwood knew better. Even though the Hellcats had maintained strict radio silence except for when it was necessary to communicate with one another over the tactical network, FRUPAC had picked up snatches of these weak ship-to-ship radio messages broadcast in the clear. That they couldn't pick up more of them was due to atmospheric conditions, Japanese jamming, and the short range at which the radios broadcast. But enough of them got through for Lockwood to assemble a mental picture of what was happening almost hour by hour.

Now, as the clock wound down on Barney, he could only sit tight and wait for a "mission accomplished" message from Earl Hydeman. According to Lockwood, the days and nights of the penetration of Tsushima had been agonizing for him and his staff. But the days and nights of operations in the Sea of Japan itself and then the breakout were infinitely worse. As he waited his thoughts once again turned to Mush Morton and the *Wahoo*.

"While it is true," Lockwood wrote, "that no single submarine nor any single crew can be considered as more important than any other in the mind and heart of a Force Commander, it is equally true that circumstances do arise wherein a ship and its men can become more closely identified with the thinking of [that commander] than other men and vessels under his jurisdiction.

"Such was true of the *Wahoo* and Mush Morton. [He] had come deeper within the orbit of my personal thinking than ordinarily happens during combat operations. [B]ecause Morton had the kind of personality that impresses itself upon people.

"War is a game you must take in your stride. There is no time for mourning or for revenge."[2] Yet Lockwood saw an image in his mind, he said, of a ghostly *Wahoo* and the face of Dudley Morton. Lockwood may have feared that he could never be at peace with himself until the two were avenged by the Hellcats.

Then came Hydeman's message: the relative ease with which they penetrated the minefields, a rough tally of the ships the Hellcats had sunk—at least twenty-seven plus a sub—their size, etc.[*] The news wasn't all good: the *Bonefish* was missing. Despite news that the *Bonefish* might have been lost, Lockwood rejoiced. Writing to his friend James Fife, Lockwood included a brief summary of the information Hydeman had radioed to Pearl Harbor.

> We had fine news from Hydeman's Hellcats last night from which it appears that all 9 got in running at depths of 130 and 150 ft. according to 2 reports and exiting through La Pérouse on surface in a fog. Bonefish *did not rendezvous for exiting*

---

*See Appendix Two for a tally of sinkings by each of the Hellcat submarines.

*although she had been talked to a few days before. I have hopes
she may still exit and there are indications she may come out of
Tsushima. I consider this proves the worth of FM Sonar and, as
you know, we are rushing installation of next 9 full speed. I
expect to find that some of those lads ran very deep and
practically disregarded FM Sonar. . . .*

*I am going to Pearl about 5 July to get all the straight dope
when the FM boats arrive there.*[3]

————

The part of the mission where the Hellcats made their exit from the Sea
of Japan had happened days before Hydeman's radio message to Lockwood. On June 23 eight Hellcats formed up in two parallel columns in the
half-light of dusk outside the western entrance to La Pérouse Strait. Their
next move would be a rush into the unknown. Before departing Guam,
Hydeman had invited review by his skippers of the provisions for Sonar
Yoke, a surface transit of the strait as set out in Operation Order 112-45.

[Sonar Yoke] Exit will be made through La Pérouse Strait
[surfaced at night on June 24]. Sunrise on 24 June is 0342, and
sunset 1924. Hence, daylight extends seventeen hours. Depending on intelligence information concerning anti-surface
mines and enemy activities in La Pérouse Strait now available
and received during Operation "Barney" [sic], Commander
Submarine Force Pacific Fleet, will recommend what appears
to be the more feasible exit plan to the Pack Commander. Exit
plans for [Sonar Yoke] are as follows.

This will be a high-speed surface dash . . . speed to be des-
ignated by Pack Commander and with ships closed up in column as much as practicable. Be prepared to furnish gun fire
support for each other if this becomes necessary.

Now, on the eve of their breakout, Hydeman had to hope that the
business about gunfire support wouldn't be necessary: Nine subs armed
with five-inch guns were no match for Japanese patrol boats. And he had

to hope that the Japanese, who by then had recovered from the shock of U.S. subs shooting up their private lake, hadn't sown surface mines to catch the raiders as they hightailed it through La Pérouse Strait.* If they had there was only one way to find out—"All ahead flank and keep your fingers crossed." As for the diversionary bombardment east of the Tsushima Strait,† it remained to be seen if it would draw Japanese forces away from La Pérouse, as Tsushima was nine hundred miles away. From what Hydeman and his Hellcats had seen so far, Japanese antisubmarine activity had been relatively weak. Therefore, he was confident that they weren't capable of stopping the Hellcats from making their escape.

These were anxious hours as the Hellcats prepared to make their high-speed dash into the Sea of Okhotsk. Gun crews stood by. Watertight doors in all compartments remained dogged so that if a mine were to blow a hole in a sub's hull, the men in other parts of the ship would at least have a chance at survival, slim though they'd be in waters with temperatures near freezing.

At the appointed hour of departure urgent radio messages to the *Bonefish* still went unanswered. Apprehension over her fate had spread among the Hellcat crews as the clock ticked down. There could be any number of reasons why she hadn't shown up—hull damage, engine failure, flooding—the possibilities were endless. Though it went unsaid, the submariners knew that there was a good possibility that the *Bonefish* had been sunk. Hydeman waited until 0300 Sonar Day for her to show up. When she didn't, he had no choice but to order the Hellcats under way. Delay meant risking discovery and attack by antisubmarine units.

George Pierce, Edge's task group commander, asked permission to stay behind to wait for her and if necessary render assistance if Edge needed it. Hydeman gave Pierce the okay to wait outside La Pérouse Strait in the Sea of Okhotsk but only for two days. If the *Bonefish* didn't show by then, Pierce was to make tracks for Pearl Harbor.

---

*Based on intelligence developed during the raid, Lockwood warned Hydeman by radio to be on the lookout for increased antisubmarine activity as well as for minelayers sent into La Pérouse Strait to sow more surface mines.
†The USS *Trutta* (SS-421) lobbed five-inch shells at Hirado Shima.

After an all-day submerged crawl from their starting points in the western approaches to La Pérouse Strait, the eight Hellcats surfaced after dark. Formed up in two columns of four ships each, they began their dash to safety, diesels roaring at full song. A heavy fog provided cover for the subs, a fog that Hydeman claimed was the only good fog he'd ever seen. Up ahead there was no way to know what they might encounter. Anything was possible, from a flotilla of enemy patrol boats to drifting mines to a convoy of plodding *marus*. The Hellcats also had to stay alert to avoid any Soviet ships that might be in the strait.

Hydeman and the other skippers would have to rely solely on radar bearings and ranges for navigation through the strait and for keeping position on the other Hellcats running in two columns two miles apart, a mile-and-a-half separation ahead and astern of one another. Their challenge was to keep from running into one another at night in fog in unfamiliar waters. Leading the way in the *Sea Dog*, Earl Hydeman's job was to see that they didn't.

La Pérouse Strait is approximately sixty miles long, depending on where a navigator plots the start of its western end and cares to mark its eastern terminus. The fifty-mile-wide open crab claw of Karafuto lies north of the strait. Its twin-taloned points, Nishi Notoro Misaki and Naka Shiretoko Misaki, form a large bay trawled by fishermen. It also provided a base for Japanese patrol boats. One of those points, Nishi Notoro Misaki, protrudes into the strait, pinching its navigable width down to less than twenty-five miles. East of this point the strait opens wide toward the northeastern coast of Hokkaido and, beyond, the Sea of Okhotsk. For the Hellcats, passage through the pinched area around Nishi Notoro Misaki posed the greatest risk. If they didn't run into any Japanese patrol boats there, it would be clear sailing all the way to the Sea of Okhotsk.

The first indication of trouble arrived when the *Sea Dog*'s radar failed— again—forcing Hydeman to slow down and drop astern of the *Skate* and relinquish the lead to the *Crevalle*. "A questionable distinction," said Steinmetz, thinking about mines. Ozzie Lynch in the *Skate* coached the radar-blind *Sea Dog* into position behind the *Skate* as the pack sped southeast on a course that would keep them away from the coast of Hokkaido.

Then, at 2245, more trouble—radar contact on a big ship due east of the Hellcats and on a course opposite their own. Was the ship Russian or Japanese? Steinmetz in the lead closed in on the contact. He saw two lights that he judged too bright for masthead lights. He decided that the ship was either a Russian freighter or a Japanese hospital ship. He peered into the gloom. Or was it a Japanese destroyer trying to sucker the Hellcats? Had it spotted them racing east low in the water? There was no way to tell. Steinmetz relayed his assessment down the line to the *Skate* and to Hydeman in the blind *Sea Dog*. Like Steinmetz, the other Hellcats watched the blip on their radar screens grow bigger and bigger as the unidentified ship approached. Still hidden by fog except for two muted running lights, the ship plodded steadily west without any sign of having spotted the band of submarine raiders, now less than a half mile distant, coming up on her port side.

All at once she loomed up out of the fog, machinery pounding away, propeller thrashing, dazzling lights haloed by swirling fog. To make certain there was no confusion over her nationality, the crew had rigged a bright light to shine on the Russian flag flying from a staff at her stern. And, as if curious about all that noise growing louder and louder off her port side—the muffled pulse of thirty-two diesel engines running wide-open—the Russians snapped on a big searchlight whose fog-streamed loom of white light they played over the Hellcats.

"Shut that goddamned thing off or I'll shut it off for you," Steinmetz raged. A gunner's mate on the *Crevalle*'s bridge caressed the triggers of a .50-caliber Browning machine gun, ready to hose down the Russian ship if Steinmetz gave the order. As if the Russians had heard Steinmetz raging, the searchlight snapped off, plunging the strait into full dark. The Hellcats jinked to the right to give the ship a wider berth, then jinked back on course. As the ship vanished into swirling fog, Operation Barney came to an end.

"It was fantastic to believe," Steinmetz later wrote in his patrol report, "that we could have gotten away with what we did against even mediocre opposition." But, he warned, "The Japs won't be napping next time, if there is a next time."

At daybreak on the twenty-fifth the *Sea Dog*, her radar back online, resumed the lead. After clearing the strait, Hydeman radioed Lockwood with an after-action report that included more information on the missing *Bonefish*. Worried that she had been damaged and unable to radio for help, Hydeman still held out hope that she might show up. He cautioned Pierce to stay alert for Japanese sub hunters while he stayed behind to wait for word from the *Bonefish*. When the forty-eight hours allotted for this by Hydeman were up, and if there was no contact, Pierce was to get under way for Pearl Harbor.

Lockwood received Hydeman's latest report with jubilation. They had done it! They had proved that subs could get in and get out of the Sea of Japan relying on FM sonar. The gadget worked. It proved too that Lockwood's plan was sound and that there was no way the Japanese could stop future raids. Their fate was sealed. All the hard work he and his staff had poured into Operation Barney had paid off. The submarine force was assured of having an important role to play in the final downfall of Japan. His submarines would finish off the Japanese by strangling them in their island bastion.

Lockwood drafted a rousing reply to Hydeman's message, which he intended should set the stage for their homecoming. They were due in Pearl Harbor for a grand reception by Lockwood and his staff. Clear of La Pérouse, seven Hellcats set their courses for Midway to refuel and take on torpedoes. Pierce waved them on as he dropped out of the pack to begin a vigil for the *Bonefish*. If Edge showed up needing help, Pierce would assist if he could.

His vigil proved lonely and fruitless. Hour after hour Pierce urged the *Tunny*'s radiomen to peak the ship's transmitter to the tactical frequency and to try again and again to raise the *Bonefish*. "*Tunny* 62V607 to *Bonefish* 67V607—62V607 to 67V607." But hissing static and the silence of the missing submarine confirmed what they'd suspected all along. When Pierce stopped by the radio room the radiomen just shook their heads. "Sorry, Captain, nothing." Reluctantly, sadly, Pierce shoved off to rejoin the Hellcats.

The Hellcats arrived in Pearl Harbor in two separate groups on the fourth and fifth of July. Though the missing *Bonefish* dampened the jubilation of their arrival, they nevertheless received a tumultuous welcome. Admiral Nimitz, a submariner himself, was there to shake hands with each member of the Hellcat crews. Lockwood and the ComSubPac staff had streamed out of their offices en masse to meet the returning Hellcats at the sub base piers. Lockwood had even rounded up another bevy of good-looking nurses to join the homecoming festivities, which included a Navy brass band playing "Anchors Aweigh" and Broadway show tunes.

Though Lockwood was a realist when it came to submarines that were overdue from patrol and presumed lost, he held out hope that the *Bonefish* might yet show up. Despite her apparent loss, he pronounced the mission a complete success, believing that Japan had been irrevocably weakened by the sinking of twenty-eight more ships and numerous small craft she could not afford to lose. He allowed that in the midst of rejoicing over Barney's success, there was no time for mourning losses or to question whether the sinking of a Hellcat submarine was a price worth paying to prove that FM sonar worked, and to avenge Mush Morton and the *Wahoo*.

Lockwood wasn't then in the mood to reflect on such matters. That would come later. He was too busy laying on a meeting in the sub base auditorium, which he filled with the officers of the sub force. Barney Sieglaff introduced the Hellcat skippers, each of whom got to tell his story, which they repeated during a formal press conference that followed the meeting between the skippers and the staff. Reporters were not informed that Edge and the *Bonefish* were missing. ComSubPac allowed that the eight skippers present at the news conference were the only ones involved in the raid. The reason for not revealing that a sub had been lost was to protect the identities of any possible *Bonefish* survivors captured by the Japanese, slim as those chances were. Speaking at the press conference, Lockwood announced that more subs would soon follow the Hellcats into the Sea of Japan. In fact, he'd already compiled a list of seven FMS-equipped boats slated for training and deployment as soon as the Hellcat skippers underwent debriefing by the Office of Naval Intelligence (ONI). As the news conference ended, and to groans of displeasure from the assembled reporters, the ComSubPac press officer announced that the Navy had embargoed the release of all news stories about Operation Barney until further notice.

After concluding the grueling business of meetings, news conferences, and debriefings, Lockwood hosted a dinner dance for the skippers. Lockwood, still the big-picture man, saw to it that each officer had a lovely female companion to keep him company. Enlisted men from the Hellcat boats made do with beer and booze on notorious Hotel Street in Honolulu. As for female companionship, they probably had the better time of it.

Soon enough it was back to work for the Hellcat submariners; there were still regular war patrols to prepare for, and, for Lockwood, more FMS training at sea. Lockwood was as busy as ever; demands on his time had not slackened a bit. Toward mid-July he had to face the fact that continuing to hope for the *Bonefish*'s return was futile. If she'd been able to get out of the Sea of Japan or radio for help, she would have by now. As for the possibilities of survivors, he was certain there weren't any. Decrypted Japanese reports of attacks on the Hellcat subs during their raid had been so garbled that no information concerning the *Bonefish* could be extracted from them by the experts at ICPOA. Her loss remained a mystery.

Lockwood had no choice but to send a letter to Admiral Nimitz that served the purpose of formally recording the *Bonefish* in the official casualty record kept by the Pacific Fleet. Lockwood titled his letter, "The USS Bonefish (SS-223)—loss of." In it he wrote, "It is with the deepest regret that I report the USS Bonefish is overdue from patrol, and must be presumed to be lost. This was the eighth patrol of the Bonefish and the fourth patrol for Commander Lawrence L. Edge. Commander Edge was an exceptionally brilliant officer, and had been awarded a Bronze Star for the Bonefish's fifth patrol and the Navy Cross for the sixth patrol. He was awarded but not presented with a Gold Star in lieu of a second Navy Cross for the Bonefish's seventh patrol. The Bonefish has a long record of many successes during her war career. She is credited with inflicting [heavy] damage upon the enemy."[4]

Admiral Nimitz forwarded Lockwood's letter to Admiral King, saying, "Forwarded with profound regret. The combat records of both Bonefish and her Commanding Officer, Commander Edge, were in accordance with the high standards of the Submarine Force, Pacific Fleet. The loss of both will be keenly felt."[5]

On July 28, one month after the Bonefish's disappearance, with Admiral King's approval, BuPers sent missing-in-action telegrams to the families of the men lost aboard the submarine. It was Navy policy to list men who had disappeared under such circumstances as missing, not dead. They would not be officially listed as killed in action until after a full year had passed from the time of their disappearance. This wording of the telegrams served to keep hopes alive among the families that at least some of the men might have been taken prisoner and that, with the war coming to an end, they might be found in POW camps. At the time of the announcement, no one, least of all ComSubPac, knew the facts surrounding the loss of the Bonefish. It would take many postwar months of searching through Japanese records to learn her fate.

An official telegram informing the families that their loved ones were missing in action was a cold, impersonal, and heartbreaking thing. It didn't tell the whole story of Edge's and his men's sacrifice; only Lockwood could do that. He noted in his diary what he had to do to rectify that situation: "Must send a letter to Mrs. Edge."[6]

Even as Lockwood made this note, events that would change the world forever began to unfold. And as they did, Operation Barney passed into history.

CHAPTER EIGHTEEN
The Long Search

Lightning erupted across the night sky. The rumble of distant thunder portended a possible break in the terrible heat and humidity gripping Atlanta. It couldn't happen soon enough for a very pregnant Sarah Edge. Though a fan moved the soupy air around in the apartment on Collier Road, the rooms still felt like ovens. Uncomfortable as Sarah was, it hadn't diminished the joy she felt over the pending birth of her and Lawrence's child, and her eagerness to share that joy with her husband. It had been two months since she'd received a letter or cable from Lawrence, longer than any period so far. A dram of fear began gnawing at her. Perhaps writing a letter to him would make it go away.

Thursday—11:10 p.m.
July 26, 1945

Dearest Sweetest Love,

Only a note tonight, because it is late again. Went to Dr. Upshaw today and think I'll stick to him for the delivery. Today he listened to the heartbeat and said, "Well, I think it's a boy!" . . . But I said, "You've been saying a girl." He said, "I say it's a boy now." Time will tell! About three more weeks to be exact!

Shug, you must come in before then, because you must have the news promptly! Gee, time is flying and no cable [from you] yet. I [contacted] W. U. again today asking if they had one for me. Guess I should stop asking and see if one comes. The office which delivers to Collier promised to phone me here first, and today the girl remembered my name and still has the note on it, so I'll just sit and wait.

I went home after I left Dr. Upshaw and looked for one stuck under the door, but no envelope! I also looked up to see the extent of your longest patrol. A few more days and this will equal it! . . .

Think I'll stop and go re-read some more of your last group of letters to me. They are such a help when I can't get new ones. Good night, dearest, and sweetest of dreams, always.

[Sarah][1]

Sarah never mailed this letter. Before she could post it she received that telegram from the Navy with its awful, heart-stopping news that Lawrence was missing in action.

Desperate for information, she had no one in the Navy to turn to for answers in her moment of agony. What kind of mission was Lawrence on when his submarine disappeared? Had the *Bonefish* been sunk by the Japanese? Or had it been an operational disaster, flooding or fire, or something equally terrible? Were there any survivors? If so, had they been captured by the Japanese? Who could answer her questions and those of the families of the men under Lawrence's command? Sarah was a woman of deep faith. Until she had answers she would have to pray with all her heart and soul that somehow Lawrence was still alive somewhere, even in a POW camp. She told herself that as long as she clung to her faith things would work out okay. She had to believe that Lawrence would return home and wrap her, Boo, and their soon-to-be-born son in his arms.

Thousands of miles away on Guam, just days before Sarah Edge received her world-changing telegram, Admiral Nimitz and his invited guests, Admiral Spruance, General Curtis LeMay of the AAF's XXI Bomber Command, and certain officers of their respective staffs, received a visit from

Captain William S. Parsons, a naval ordnance expert assigned to the Manhattan Project. He had come to brief the officers on recent developments concerning the atomic bomb.

A lot had happened since the first briefing Nimitz had received in February. The U.S. Third Fleet had about completed its destruction of the Imperial Japanese Navy; LeMay's B-29s had reduced vast areas of Japan's major cities to ashes; Japan, seeking a possible end to the war on terms less onerous than unconditional surrender, had sent nascent peace feelers to the Allies through its ambassador to the Soviets; and, while planning continued for Operation Olympic, the invasion of Kyushu, Washington prepared for a possible Japanese surrender by assembling staffs of experts in the fields of civil affairs, reconstruction, and, most important, war crimes. The issue of Japan's extending peace feelers seemed to indicate that the empire had reached the end of the road. Still, just as before, no one thought the final collapse would come without a bloody fight on the Tokyo Plain.

Parsons screened a color motion picture of the successful test of the world's first atomic bomb at Alamogordo on July 16. To a man the officers found it a sobering, if not frightening spectacle. It didn't take much imagination to picture what such a weapon would do to a city and its population, and it changed some minds regarding its usefulness as a weapon to force Japan to capitulate. One of those minds may have been Nimitz's. He'd once told King that he believed America's objectives in the Pacific would more likely be realized by continuing to blockade Japan and destroying her military forces than by relying on a secret weapon. But it was clear to him now that the sheer visual effects produced by the bomb, to say nothing of its destructive power, would shock and demoralize the Japanese people and sap their resolve to fight on.

The officers also learned from Parsons that the heavy cruiser USS *Indianapolis* (CA-35) was on her way to deliver the subcritical core of U-235 to the scientists on Tinian assembling the atomic bomb, which, if President Truman gave the order, would be dropped by B-29 on Japan. Washington hoped that with the announcement of the Potsdam Declaration, setting forth the terms for Japan's surrender, Tokyo would agree to end the war. But the refusal of Japan's leaders to face reality had deflated those hopes.

Nimitz, frustrated over Japan's intransigence, believed that, atomic bomb or no atomic bomb, it was just a matter of time before they caved in. But how much time and how many more lives it would take were questions neither Nimitz nor anyone else could answer. And until the Japanese did cave in, Spruance would continue launching air attacks, LeMay would keep dropping incendiaries, and Lockwood would keep on sending subs into the Sea of Japan to mop up *marus*.

Within ComSubPac operations at Guam, as in every other command, there was a growing belief that the end was near. Aboard the last wave of seven FMS subs already headed for the Sea of Japan the sense of urgency was palpable; the submariners wanted to get in their last licks before the war ended.

Edward L. Beach, skipper of one of those boats, the USS *Piper* (SS-409), felt there was a good possibility he'd be too late.[2] "One decision I made," wrote Beach, "and clung to tenaciously: we were going to get *Piper* into action or break our necks trying." Beach described this overwhelming need for action when he wrote, "[We] raced for the war zone. Somehow I felt it was slipping away from us—receding faster than we approached it. . . . I felt an overwhelming impatience to be back in it before it ended." And, Beach added, now that the Japanese were near defeat, he wanted to destroy what was left of them just as they had destroyed the *Wahoo* and so many other subs.

As Beach and the *Piper* raced for the Sea of Japan, the war itself raced to its conclusion. On August 6 an atomic bomb devastated Hiroshima. On August 9 another atomic bomb destroyed Nagasaki.

While a world stunned by the use of this frightening new weapon waited for news of Japan's capitulation, Admiral Lockwood officially announced the loss of the *Bonefish*. Thus, a notice appearing in the August 12 edition of the *Atlanta Journal* announced the end of one life and the beginning of another.

SON BORN DAY SKIPPER OF SUB ANNOUNCED LOST

The Navy Department Saturday announced in Washington the loss of the submarine Bonefish on the same day [August 11] that a son was born in Atlanta to the wife of the vessel's skipper, Commander Lawrence L. Edge.

Mrs. Edge was notified two weeks ago that her husband . . . is missing in action. The Navy Department said the Bonefish is "overdue from patrol and presumed lost."

The boy has been named Lawrence Lott Edge, Jr.[3]

By coincidence an article about American POWs appeared in the same edition of that newspaper. The POW story surely must have strengthened Sarah's resolve to believe that Lawrence was alive and in enemy hands. "Jap Surrender to Free 16,700 U.S. Prisoners,"[4] headlined the article reporting that American POWs were incarcerated in prisons across Japan and in occupied territories. The numbers were incomplete because it had been impossible for the Red Cross and other neutrals to visit the camps during the war to assess the situation. It was known from official records, the article added, that there were over two thousand naval personnel alone being held. For Sarah and the families of the *Bonefish* crew, it meant that there was indeed a good chance that Lawrence and his men might be among them.

Following Lockwood's announcement about the *Bonefish*, the Navy lifted its embargo on news about Operation Barney. Within hours of its lifting the *Los Angeles Times* published a long article entitled "Sub Flotilla Returns After Taking Nip Toll." Written by Kyle Palmer, *LA Times* war correspondent, the story provided a reasonably accurate account of Operation Barney, its objectives, and its results, including the number of enemy ships sunk and damaged. In his article Palmer went on to say, "Not a man nor a ship of the striking force was lost in the operation—one of the most daring and spectacularly successful of the Pacific war."[5] Palmer and the other reporters who had attended the Hellcats news conference in Pearl Harbor were apparently still unaware at the time of release and publication of the story that the *Bonefish* had been lost during the raid.

Sarah Edge and the families of the missing *Bonefish* crew still didn't know that the *Bonefish* had been lost in Operation Barney. Sarah learned that her husband had been in the Sea of Japan only when she received a letter from Lockwood fulfilling his pledge to write her. The letter, which arrived while Sarah was still in the hospital after the birth of her son, provided information about Operation Barney and, based on sketchy details then available to ComSubPac, Lockwood's explanation of how Lawrence and his crew had likely perished. Meant to bolster Sarah's spirits, the letter only raised more questions, and, later, invited controversy and criticism by *Bonefish* families regarding Lockwood's judgment in the matter.

12 August, 1945

My dear Mrs. Edge [Lockwood wrote]:

In the midst of national rejoicing at the probable end of the war, it is particularly painful that you should be informed that your husband, Commander Lawrence L. Edge, U.S. Navy, is missing.

Undoubtedly the Navy Department has already informed you, since today's papers contain the news that the Bonefish has been declared missing and must be presumed lost.

I cannot give you entire details for reasons of security, but as you may know, Larry, as we all called him, had special gear in his ship, the Bonefish, which made it possible for him to join in a raid into the Japan Sea. They got in without a hitch about 5 June and Larry was talked to by Commander George Pierce, Captain of the Tunny, on 18 June. The Bonefish reported that she had sunk two ships and was proceeding in to Toyama Wan, a bay on the west coast of Honshu.

That is the last time that Larry's ship was seen, and

*although after exit on the day set for the raiding submariners to
depart that area, Pierce stayed just outside for two days trying
to contact the Bonefish by radio; no message has ever been
received from her. I also sent her two messages advising her of
the best means of making exit, but these were never answered.*

*What happened to her we do not know, but so far we have
no information that the enemy has publicly claimed the sinking
or capture.*

*I earnestly pray that Larry did survive her loss, but knowing
how few submarine personnel have been reported as prisoners,
I cannot conscientiously encourage you to believe he did. . . .
Please accept my deepest sympathy in your sorrow, and know
that we all pray that Larry may have survived the loss of his
ship.*

Sincerely
C. A. Lockwood, Jr. [signed][6]

Shortly after receiving Lockwood's letter, Sarah received one from
Lawrence's division commander, Captain Lucius Chappel. After concur-
ring with Lockwood's description of the raid, Chappel sounded a hopeful note.

*I beg you not to utterly despair. In the waters in which the
Bonefish was operating it is [quite] possible that some of the
ship's company survived. Many submarine people have turned
up, months or years after they were declared missing, in
prisoner of war camps. My old ship is an example [USS* Sculpin
*(SS-191)]; lost early in 1944, yet it was not until this spring that
we learned that a good part of the crew was safe and well.*[7]

Confined to a maternity ward after the birth of Lawrence Jr., Sarah felt
helpless. There was so much that she didn't know and wasn't being told.
Lacking official information and ignorant of the facts surrounding
Operation Barney except for what she'd learned from Lockwood and
Chappel and from what had been published in the paper, she was desper-

ate to know what had happened to the *Bonefish*. Adding to her confusion were a host of unanswered questions. For example, given that the war was almost over, why had it been necessary for the *Bonefish* and the other submarines to penetrate the Sea of Japan? What, if anything, aside from proving how audacious and daring submariners were, had it accomplished? Was the mission supposed to have ended the war? If so, it had failed; it was almost mid-August and the Japanese still hadn't surrendered even after being hit by two atomic bombs. How, then, could a small task force of submarines possibly bring about the end the war? Most important of all, was there a real possibility that there may have been survivors from the *Bonefish* who were being held in Japanese POW camps? Chappel seemed to think there might be, while Lockwood seemed to hold out little hope for their survival. Given all the uncertainties, Sarah pledged that she'd never abandon hope for Lawrence and his men until there was incontrovertible evidence that they were dead, not until every last Japanese POW camp was liberated and every man in them identified. Lucius Chappel had begged her not to despair, said that it was possible that they had survived. Sarah desperately wanted to believe that that was true.

Admiral Nimitz had been counting the days since the bombings of Hiroshima and Nagasaki when he received a coded message from Admiral King. The message began, "This is a peace warning."[8] King's message stated that though it was not yet official, the Japanese had informed the United States government through the Swiss that they were willing to accept the terms of surrender contained in the Potsdam Declaration. While Washington waited for an official announcement from Tokyo, the Navy continued to launch air attacks across Honshu. On Tinian, Curtis LeMay's B-29s stoked up on incendiaries to drop on Japanese cities.

On August fifteenth, at CinCPac headquarters on Guam (the fourteenth in Washington), a restricted-use eyes-only Teletype began spewing out a message in plain English, not code. Nimitz's intel officer tore the top sheet from the machine, read Admiral King's message, then dashed for his boss's office.

Nimitz looked up, surprised. "It's over," blurted the excited intel officer. Nimitz read the message, which included the full text of Japan's acceptance of surrender transmitted via the Swiss legation in Tokyo to Washington. Nimitz didn't get excited; he just gave the officer a smile of satisfaction. He'd known all along that the Japanese were finished, and now at last it *was* over. Nimitz wasted no time broadcasting an immediate cease-fire order to the fleet, which he followed with a personal message to sailors across the Pacific on the ending of the war. Lockwood broadcast a message of his own to the submarine force. It read in part:

> The long-awaited day has come and cease fire has been sounded. As Force Commander I desire to congratulate each and every officer and man of the Submarine Force upon a job superbly well done. My admiration for your daring, skill, initiative, determination and loyalty cannot be adequately expressed. . . . You have deserved the lasting peace which we all hope has been won for future generations. . . . May God rest the gallant souls of those missing presumed lost.

———

The end of the war brought a push for demobilization and, with it, cries to get the troops home by Christmas. Lockwood was caught up in a whirl-wind of meetings and conferences designed to do just that, piled on top of the Everest of paperwork generated by the cessation of hostilities. He hardly had time to reflect on what his sub force had accomplished and at what cost. He was thankful the war was over—thankful, too, that he'd not have to send any more young men on patrols they might not survive. There were, given the small size of the force, far too many who did not return.

As for Operation Barney, Lockwood believed it pointed the way to the future of the submarine force through operations that would capital-ize on ever more sophisticated submarines and weapons. Not everyone shared Lockwood's view. Critics within the sub force observed that de-spite the bravery and dedication of the submariners, the results of the recent forays into the Sea of Japan hardly seemed worth the risk, espe-cially in regard to the loss of the *Bonefish*, whose fate was yet to be determined. Critics outside the sub force believed that Barney had been a stunt, a make-work patrol, designed by Lockwood to keep his sub force occupied as the war wound down. They pointed out that no matter how many ships the Hellcats sank in the Sea of Japan (the seven submarines that followed the Hellcats into the sea in August 1945 sank only two ships), it had had no measurable effect on the weakening of Japan's ability to prolong the war. And though the sub force had almost single-handedly destroyed Japan's merchant marine and a good portion of her navy, the force nevertheless stood to be eclipsed by Army and Navy air power. In the end the atomic bomb proved to be the decisive weapon that ended the war in the Pacific, not sea power. By comparison Operation Barney was just a relatively minor affair. Lockwood, always an optimist, shrugged off the criticism as he heaped deserved praise on his submariners and looked to the future.

For the moment, however, the future of the silent service looked bleak. Occupied with the logistics of winding down the robust wartime sub force, Lockwood saw a shrinking fleet of mostly battle-weary subs and submariners. He knew America would need a new and modern fleet of long-range, high-endurance submarines and that such a fleet would require a complete rethinking of what a submarine should be.

While grappling with these issues, Lockwood flew with Admiral Nimitz and a group of other officers to Tokyo Bay, where they participated in the surrender ceremony held aboard the USS *Missouri* (BB-63) on September 2, 1945. After the ceremony Lockwood took the opportunity to tour several captured Japanese submarines capable of launching aircraft from waterproof hangars on their decks. Among them was the *I-400*, at that time one of the biggest submarines ever built.

Lockwood was shocked when he saw these submarines. Their sheer size alone staggered the imagination. The Japanese boats were far more advanced than Lockwood could have envisioned. The design of their diesels incorporated up-to-date technologies; their radar and sonar systems were on a par with those of the U.S. Navy. He also got a close look at the fabled Japanese "Long Lance" torpedo, which vastly outperformed by a whopping margin any torpedoes the U.S. Navy had in its inventory.

Though Lockwood was dismissive of the big subs for their lack of refinement and creature comforts in comparison to U.S. subs, clearly their great size pointed to possibilities for the future of submarine development.

Lockwood's return to Guam coincided with the arrival of some of the first American POWs from camps liberated in Japan. Many of the survivors (only a few submariners were among the returnees, none from the *Bonefish*) were in terrible condition, suffering from dysentery, jaundice, malnutrition, and physical abuse. Far too many showed evidence of the brutality and the inhuman conditions they had suffered in captivity. Many of the returnees had to be hospitalized; those in better condition, and able to withstand it, were debriefed by naval intelligence to gather information for use in the upcoming war crimes trials.

It was a pitiful, embittering sight, Lockwood said, to see those men, their skeletal, hard-planed, sunken-eyed faces haunted by what they'd experienced. He was horrified to learn what the Japanese had done to them and to learn that the Japanese had murdered and maimed prisoners as they saw fit. How many submariners had survived the sinking of their ships only to succumb to torture and disease would not be known for months. Meanwhile, Lockwood's staff kept a tally of submariners returning from POW camps, hoping to get information from them about the fate of missing shipmates. Among the first returnees to Guam, not one of them had seen or heard of any survivors from the *Bonefish*.

Home again and with a new baby to care for along with Boo, Sarah began to assemble all the information she could find about the loss of the *Bonefish*. What she had so far were just the basic facts provided by Lockwood concerning a raid by submarines in the Sea of Japan. They included a couple of names, a date, and a bay in Honshu called Toyama Wan. This wasn't enough. She had to know more. Somehow she had to get every piece of information available. If she could, it might explain what had happened to the *Bonefish* and help her cope with the fact that Lawrence was missing in action, perhaps dead. The *Bonefish* families, coping with their own losses, were no better off. They knew almost nothing regarding the fate of their loved ones beyond what they'd been told by the Navy, which wasn't much. In distress they turned to Sarah for help.*

> *Mrs. Edge, our hearts go out to you in your great sorrow [wrote the sister of a Bonefish sailor] and may God bless and comfort you. In the paper your husband's name was given as commander of the sub. His name is the only one we have of anyone that was with [my brother]. It seems so terrible that they could not have made this patrol safely. It might have been their last in those dangerous waters, as the war news [was] so good and encouraging.*
>
> *Mrs. Edge do you know or have any idea of where the Bonefish was? [Y]ou don't know how very much we would appreciate it, if you would tell us the least bit of news about the location of the ship, and if you have learned any more than we have from [the Navy]. Do you hold the least bit of hope for them? That's an awful blunt question, I know, but to get the opinion of someone who has a great interest in the same ship seems like [it] would help. We hold very little hope but seems like we must try to.*

The mother of a quartermaster serving in the *Bonefish* wrote:

> *[I] wonder now if you know any more about it than we do and are you hoping or knowing that they were picked up or even taken prisoners. [sic]*

The mother of another *Bonefish* sailor had reason to be hopeful.

> *Saw in last night's . . . paper, a family . . . just received a wire from Red Cross that their son is alive and well—was found*

---

*This and other extracts from the letters that follow are from some of the hundreds written by the families of *Bonefish* personnel to Sarah Edge over a period of approximately two years. This trove of letters, valuable documents in their own right, is part of the Edge family archive.

in a Jap prison camp. The family had previously received a wire from Navy Dept.—first, that he was missing and a later and final wire that he was dead—so miracles do happen. [A friend on Guam] thinks there is a good possibility that officers and men may have been taken prisoners. . . . [sic]

The writer shortly sent Sarah another letter. It contained information about 156 men, survivors of the sinking of the destroyer USS *Pope* (DD-225), who had been located in a prison camp. "It evidently seems they are still locating prisoners," she said. Another writer refused to allow "three brutal words, 'missing in action,' to shatter [my] world."

Sarah received many letters praising Lawrence as a wonderful, caring skipper who would do everything in his power to bring his men home. One sailor had written home that Edge was like a father to him. Mothers, fathers, and wives told Sarah that they believed all the men would come back as soon as they were found in one of the camps. It would just be a matter of time, they said, until their prayers were answered.

Some letters contained information and news clippings detailing the return of survivors of submarines. One letter writer wrote, "[P]apers came out with a story about a lot of men being found in a Jap prison camp from the S-44, Sculpin, Tullibee, Perch, Tang, and Grenadier." The writer enclosed a clipping of a story published in *The New York Times* in late August 1945, which said, "7 Jap Ships Bagged Before Tang Sank." As explained earlier, the *Tang* had been sunk by one of her own circular-running torpedoes. Her skipper, Commander Richard O'Kane, and eight men survived the sinking but were held captive by the Japanese under extremely brutal conditions. O'Kane was lucky to have survived his ordeal. Stories like these helped ease the crush of anxiety and grief bearing down on the families of the missing *Bonefish* men. And they kept hopes alive that eventually they'd be recovered.

While Sarah was digging for information and answering letters from the families, false eyewitness accounts and rumors about the sinking of the *Bonefish* began circulating. These cruel, deliberately fabricated stories concocted by sailors with limited knowledge of submarine operations were replete with erroneous information and impossible scenarios. For the anguished families it was almost more than they could bear. The men who made these false claims were eventually exposed and punished by the Navy.

A relative of a *Bonefish* sailor, after hearing these rumors, wrote Sarah that a friend had spoken to several submariners who claimed to have seen the *Bonefish* attack a large convoy, after which she was depth-charged by destroyers and sunk. Another story circulated that the *Bonefish* was last heard from going into Toyama Wan and that she had been attacked and sunk by midget subs. Both stories were false. No one ever saw the *Bonefish* attack a large convoy in the Sea of Japan, and the Japanese had no active midget subs deployed in the western Pacific.

Another false rumor claimed that the *Bonefish*, plagued by radio problems and unable to communicate, was hiding out somewhere off the coast of Manchuria or Siberia. Yet another said that her crew had sought refuge on one of the tiny volcanic islands dotting the Sea of Japan and, not knowing that the war was over, were awaiting rescue. Another claimed that part of the crew had been taken prisoner and that they were safe. Some of these rumors gained credence because the families grasped at anything that might offer hope for the men's survival. While most rumors got started through ignorance or from a lack of knowledge of submarine operations, others were started deliberately.

The cruelest rumor of all got started by a submariner who sent a letter to the mother of a *Bonefish* sailor. He claimed to have been only five hundred yards away from the *Bonefish* when she was torpedoed and sunk, presumably by a Japanese submarine. He also claimed to have seen the woman's son a split second before the torpedo hit, which, he said, made him the last person to see her son alive. His claim was an outright lie. So was the story told by a young submariner to the sister of a *Bonefish* sailor that the *Bonefish* had had her conning tower shot off at Iwo Jima during her seventh patrol, but that she'd managed to return to Guam. It was unthinkable, said the missing man's sister, that anyone would send men on such a dangerous mission in a ship that had been so badly damaged.

"['I was told,' she wrote] how badly the ship was damaged on her patrol before the last. And how my little brother hated to go on this last patrol, [as] he knew he would never be back. I can't understand why, if the job they were assigned to do was so dangerous, they didn't let the boys come home [on leave]. But there are a lot of things I don't understand about the Navy." Later, she wrote to Sarah that she hadn't been taken in by the sailor's wild story.

The days turned into weeks of unrelenting agony over the unknown fate of the missing men. More reports came in every day about the release of POWs, some hopeful, some not. BuPers released a statement that every effort was being made to liberate prisoners and to locate and clarify the status of personnel listed as missing. Admiral King had directed that all islands and areas not previously explored were to be subjected to a thorough and exhaustive search for U.S. and Allied personnel. It was a huge task that would take a lot of time and require the transport of hundreds of search teams over a vast area of the Pacific. Meanwhile, other teams were fanning out over the home islands of Japan, searching for prisoners in every town, village, and farm, as the Japanese, in addition to holding POWs in large, squalid camps, had also scattered POWs across the country to prevent them from being rescued.

As news reached the public about the horrible physical condition of returning prisoners, along with stories about how barbaric the Japanese treatment of captives had been, the families of the missing *Bonefish* men feared the worst.

In the absence of information that would either confirm or dispel the families' worst fears, new rumors began to spread, some darkened by anger arising from the need to blame someone other than the Japanese for the loss of the *Bonefish*. Fueled by the belief that the *Bonefish* had been sacrificed unnecessarily, that anger boiled over into the open.

The mother of a *Bonefish* officer wrote, "There is no logical person other than Admiral Lockwood to have given orders for [the *Bonefish*] to have gone on into Toyama Bay—I, personally, would like to ask Admiral L just one question, 'Would you have ordered the Bonefish on into the Bay if you had had a son a member of the crew?'" And from another family member, "A report from a boy who was on a sub tender . . . [said] that he knew [the *Bonefish*] was given extra assignment <u>over and beyond</u> [emphasis in original] other boats." "Seems to me that those 'high up' in submarine service could elicit information from Japs. I heard on the radio . . . soon after the [*Bonefish*] must have been in trouble, the Japs claim that they had sunk 6 American subs just off the mainland. Since [the *Bonefish*] seems to be the only one not accounted for I feel sure Japs stationed in Toyama areas know what happened."

A rumor spread by persons unfamiliar with how submarine war patrols were conducted suggested that somehow Lawrence and the other Hellcat skippers had requested permission from Lockwood to make the Operation Barney patrol into the Sea of Japan, and that he'd approved their request.

To put an end to such rumors, Sarah sent letters to the families in which she included copies of her correspondence with Lockwood. The families responded with a flurry of letters of appreciation for her efforts. While the information she provided didn't dispel the families' uncertainty, it helped to clarify some of the murky details and half-truths that had grown up around the missing *Bonefish*.

In mid-September Sarah received a reply to a letter she sent to Lockwood seeking answers to two important questions. Lockwood was busy administering a sub force already beginning to put itself up in mothballs, Barney far from his mind until Sarah refocused his attention on it. Her questions concerned submarine POW returnees and also whether or not Lawrence had gone on the mission into the Sea of Japan at his own request, as had been rumored. Lockwood replied to her first question but didn't answer her second question directly. Instead he forwarded a copy of her letter to George Pierce, the *Tunny*'s CO during Operation Barney.[9]

Dear Mrs. Edge [Lockwood wrote],

I have talked to a great number of our prisoners-of-war who have been released from Japanese prison camps and I regret to inform you that none know anything of any prisoners from the Bonefish, however it is possible that if any were taken they might have been held in some local camp and had not yet been put into the main camps.

*As to the letter you have received regarding the report that
the Bonefish went on this special mission at her own request, I
am sending a copy of this letter to Commander George E.
Pierce, Captain of the Tunny, who was the last person, so far as
we know, to talk with the Bonefish. Commander Pierce
reported to me that Bonefish had requested permission to enter
Toyama Wan as he [Edge] thought he might find good hunting
there. This Bay is, of course, close in to the Japanese mainland
but it is not excessively hazardous in the sense of being shallow
water, for actually the water there is very deep.*

*I will ask Commander Pierce to give you all possible
information as to his last meeting with the Bonefish.*

Sincerely
C. A. Lockwood, Jr. [signed][10]

This was the breakthrough Sarah had hoped for—contact with someone who had actually spoken to Lawrence on the scene just before the
Bonefish disappeared. The news about returning POWs was only mildly
hopeful at best, but Sarah knew that the Navy didn't declare a man dead
until he had been missing for a year, and there was still plenty of time left
to find Lawrence and the others. U.S. search teams had only just begun
their searches throughout Japan for prisoners and it would take time to
reach the camps on that country's west coast, much less the ones they
didn't know about. With her unshakable faith, Sarah believed anything
was possible, perhaps even that Lawrence and his men would return
home.

CHAPTER NINETEEN

The Hour of Sacrifice

U.S.S. TUNNY (SS-282)

Dear Mrs. Edge [wrote George Pierce]—

*I'm sorry to have taken so long in answering your letter
[forwarded by Lockwood]. I will do my best to answer your
questions.*

*. . . The boats were divided into 3 groups of three, with the
Bonefish in my group. We talked to each other on the 16th of
June and patrolled together on the 17th. On the 18th Larry
asked me for permission to enter Toyama Wan to patrol
submerged in there on the 19th. . . . I told Larry to go ahead but
to be at a certain rendezvous point by the 21st. . . . Larry stated
that he intended to stay in deep water. The depth of the
water . . . is 350 fathoms, which would preclude any salvage
operations.*

The entire operation was purely on a volunteer basis.*
*Larry's decision to enter Toyama Wan was sound and I felt
justified in allowing him to do so. I haven't even tried to guess
what happened to him. I seriously doubt that the Bonefish
struck a mine as the water he was in is rather deep for mining.
Whatever happened occurred on the 19th or 20th of June.*

---

*Pierce's "volunteer basis" apparently refers to Operation Barney itself.

Much as I hate to say it to you, capture by the Japs seems a
remote possibility. . . .

Sincerely,
George Pierce [signed][1]

For Sarah, struggling to keep faith that Lawrence would return, Pierce's assessment that capture seemed remote because the men had gone down with the *Bonefish* dealt this hope a crushing blow. Pierce described what he and Lawrence had discussed prior to Lawrence's patrol in Toyama Wan, but Pierce didn't know what had happened to the *Bonefish* any more than Lockwood did. Nor did he and Lockwood reveal what Operation Barney had accomplished. Could the mission have been so vital to the winning of a war, a war virtually won, that eighty-five men aboard the *Bonefish* had to die? It must have seemed incomprehensible to Sarah that Lawrence and the men under his command had perished little more than eight weeks before the war ended.

By mid-September U.S. servicemen were starting to stream home from Europe and the Pacific. Meanwhile, in Japan, teams of U.S. Navy intelligence experts began combing through the tens of thousands of pages of Japanese naval records, many of them handwritten, seized at the naval ministry in Tokyo. The teams were searching for technical, operational, and scientific information related to methods the Japanese armed forces had employed during the war. Other teams of specialists were searching the records for information on the fate of missing men and ships, trying to discover the circumstances of their loss. They were also trying to determine the fate of American naval personnel—submariners in particular—thought to have been captured by the Japanese navy but not yet located in POW camps.

In general, the records thought to contain this information consisted of Tabular Records of Movement, that is, the day-to-day records kept by each Japanese warship of its operations. These records provided the search teams with everything from the names of the ships to their duties, battle engagements, expenditure of ammunition, and battle damage. In addition, so-called action reports and war diaries similar to a ship's log provided the search teams with detailed accounts of an individual ship's battle actions. Though these records were in disarray, the Japanese-speaking Navy specialists began to assemble a mosaic of Japanese naval operations throughout the Pacific theater, which they broke down into smaller segments for study. At the behest of ComSubPac, a team had been detailed to identify Japanese ships that had attacked U.S. submarines. Making sense of it all was hard, slow work. Little by little a picture of Japanese antisubmarine operations began to emerge from the blizzard of paper that provided information on the fate of some missing U.S. subs. Lockwood knew, for instance, that while the Japanese had had little trouble locating submerged submarines via sonar, they consistently failed to destroy them due to a lack of persistence and skill. As a result they were quick to accept even the flimsiest evidence as proof of a kill. As the data began to trickle in from Japan Lockwood was astonished to discover that the Japanese believed that they had sunk a total of 468 U.S. submarines, far more than the total number in the entire U.S. fleet!

The search team often found itself hampered by incomplete or contradictory data. And because the work required meticulous attention to detail, it took longer than expected to piece together bits of information pointing to a specific cause for the loss of individual submarines. The team organized the reports of antisubmarine attacks by date and geographic location, then correlated that data to known information about missing U.S. subs with names like *Amberjack*, *Grampus*, *Scamp*, and *Golet*. Reports of attacks on other missing submarines, including the *Bonefish*, weren't found during this initial search.

Back in Atlanta, Sarah Edge, undaunted in her quest for answers to the *Bonefish*'s disappearance, compiled a list of questions based on the information she had painstakingly assembled from her sources.[2] A question that emerged from this information, and which troubled her and some of the families, was whether or not Lockwood had pulled some of the Operation Barney subs off patrols to send them into the Sea of Japan without giving their crews enough rest. Sarah feared that fatigue may have played a role in the loss of the *Bonefish*, given that Lawrence had written about the quick turnaround between his seventh and eighth patrols that had cut short his rest period on Guam. Another troubling question was, Had the men been ordered on the raid by Lockwood?

Sarah posed these questions and others in a letter she mailed in early October to Lawrence's friend and pack mate Ozzie Lynch, skipper of the *Skate*. Lynch had filled the electronics job that Lawrence was slated to fill after completing Operation Barney. Sarah's letter was blunt and to the point. Her words expressed all the anguish and pain caused by Lawrence's disappearance, all the unknowns and unanswered questions that had made her spirits sometimes soar and sometimes crash as she waited for news of his fate.

Dear Ozzie,

Knowing that you were not only on the same raid with Lawrence, but also in the same group of three, I am writing to ask you to please tell me all that you know concerning the loss of the Bonefish. Since there is no longer any censorship I would appreciate your being frank with me.

There's a good deal I know already but there are several points which do not make sense to me. . . . How were you ordered on the June raid? Were you given an opportunity to refuse to take your boat? Why was Lawrence given such a short rest period after his terribly dangerous [seventh] patrol which was physically exhausting to him and all the men? From his letters I gathered that this patrol was extremely dangerous and that he might not return from this one and even know whether he would have a new son or daughter. From about the middle of June until [that] horrible wire reached me, as I wrote each letter I could not help wondering if he would ever receive any of them. I find that some boys who returned from the same raid had definitely warned their families that they might not return.

Ozzie, which boats saw the ships going around [the Noto

*Peninsula] into Toyama Bay? And why was Lawrence sent in
after them? Why did he decide to go submerged at a deep depth
all the way if not to avoid mines. . . . Did you and George
[Pierce] too feel that there were mines in the waters below 100
fathoms?*

*Did you go into any bays? I know some of the boats did! I
understand Toyama Wan was more hazardous than the others.
Do you think Lawrence just sank the two ships or more? How
lucky were you?*

*Do you not believe with me that he struck a mine? Everyone
seems to feel that he was lost around or in the bay. Am I correct
in thinking that he could have been lost anywhere between the
point [where] George last contacted him, 50 miles NW of [the
peninsula] and La Pérouse Strait.*

*I still can't believe he is gone. It is too much to ask of one
right at the end of the war. Sarah keeps insisting that Daddy is
coming home "tomorrow" to see "Little Brother." I suppose you
know that Lawrence was to have left sub duty and gone to radio
work after his April patrol. He was informed that arriving at
Pearl after he left Calif. but Adm. Lockwood told him he would
have to make one more [patrol]. I will admit to you that he was
indeed disappointed.*

*. . . If you can spare a moment . . . I would appreciate your
writing me to help clear my dizzy head.*

Sincerely,
Sarah [signed][3]

Lynch answered promptly.[4] He told Sarah that the *Bonefish* had suffered no serious damage (as rumors had claimed) and was in good shape
when she shoved off on Operation Barney. "I talked to Larry quite a bit
before we started out and he seemed to feel fine about going in." As for
being ordered to take on the mission, "We were ordered to make the June
raid the same as we were ordered to make any other patrol except we had
all the special charts and gear to make the run."

"I was not asked particularly whether I wanted to make the run or not.

In my case I wanted to go. I was married once, and there was no great loss to anyone should I not return. It has always been the policy for any skipper who wanted to give up command to merely say so and he would be relieved."

As for the rest period between patrols, Lynch said, "Larry had a short rest period because he had to make the schedule with the rest of us. The operation as you may have guessed had been planned for a long time." "You are correct about the patrols being dangerous. I felt that I would not get back. . . . I did not tell anyone about it, but it's the case."

Lynch explained his role in the Toyama Wan incursion, writing, "I was the boat that saw the [Japanese ships] going into Toyama Bay and so reported it to the pack commander, George Pierce. Larry went in after them because that was his assigned patrol area. I was assigned half of it for the first two days only. . . . I do not . . . think that there were any mines below 100 fathoms.

". . . The only word we had from Larry on how many [ships] he sank was four days before we thought he was missing. He could easily have sunk more." To the hazards of Toyama Wan, Lynch said, ". . . [it] was no more hazardous than any other place except it was a bay and as such had only one direction to go to get out.

". . . You asked me how lucky I was. We were very lucky [to have sunk the ships we did.]

"No one knows just what happened to the Bonefish. We thought that some information might come from the Japanese after the war was over as to claims made by their anti-submarine vessels. There is one third of a chance that they struck a mine. There is just no way to tell just where the boat might have been sunk."

A few weeks later Ozzie Lynch received Sarah's response.

Somehow your letter seemed to lift a weight from me, especially the one sentence: "Larry went in after them because that was his assigned patrol area." So many of the families have heard that Lawrence asked to go into the bay, and I believe several felt as one mother said to me over long distance: "Why should he have asked to risk the boat and men to go into such a dangerous

*place?" No one knows better than I, that Lawrence would
never risk his boat and men anywhere he was not ordered to go
or felt it his duty to go. He personally was too anxious to survive
this war.*

*. . . [I]n the last week I have received other reports that do
not coincide with what I had figured to be true. . . . I wonder if
you think, why bother with these details. It is over and done
with but somehow it helps a small amount to know all the true
facts surrounding the loss of the boat and our men. It is
constantly in my mind, as in others, so I guess it is best to work
it off this way rather than just to sit and wait for the final
verdict from the Navy.*

In a follow-up letter to Lynch, Sarah asked for his views on a number of issues circulating among the *Bonefish* families. Foremost were the rumors and speculation that some of the men had survived the sinking and were alive in prison camps awaiting rescue. Such rumors and speculation had been stoked by a report from a Navy chaplain, the father of a *Bonefish* sailor, assigned to POW recovery in Japan. He reported that many of the POW camps had still not been reached by U.S. forces, especially those on the west coast of Japan. This was due in part to heavily mined waters in the Sea of Japan that had not yet been cleared, plus the fact that there were few roads into the areas where those camps had been built. The chaplain explained that while records were still being searched for information, none had yet been found that would indicate *Bonefish* survivors were in one of the camps. Still, the chaplain said there was reason to be hopeful.

Another troubling issue concerned those who had all but given up hope for their loved ones and sought to assess blame. For them, Lockwood had become a convenient scapegoat. It was said that he and others in authority, realizing that it looked bad for eight undamaged subs not to go back into the Sea of Japan to the aid of the *Bonefish*, had told Pierce and Lynch what they could and could not say about it.

Through dogged persistence Sarah had assembled an impressive array of facts pertaining to Operation Barney, which she included in her letters to Lynch and the families. As the wall of secrecy surrounding the operation slowly opened wider, she at last understood the mission's tactical and strategic objectives and also how the mission itself had been carried out. However, at this stage she still had no knowledge of Lockwood's desire to exact retribution, if not a measure of personal vengeance, from the Japanese for the death of Mush Morton and the loss of the *Wahoo*.

Addressing the questions Sarah posed in her follow-up letter, Lynch wrote in mid-November, "Let there be no doubt that Larry was as good a fighter as any man who ever went to sea. None of us know the story of the end of his ship, but I know that he went down fighting for you and me and the rest of us."

Most of the information that she had was correct, he said, in regard to the mission itself, the number of subs, and its dates. What was not true, he said, was that the *Bonefish* had rendezvoused with the other eight Hellcats, after which she had somehow disappeared or turned back during the escape phase. It was also not true, according to rumors, that the ship had been scuttled or that she'd been seen attacking a large convoy and was in turn attacked herself. Once again, he told Sarah that he doubted the *Bonefish* had struck a mine, as the water in which she was operating was too deep for mining. "Much as I hate to say it to you, capture by the Japs seems a remote possibility."

Sometime in the fall of 1945, Sarah sent a long letter to all of the *Bonefish* families. She told them what she had learned about Operation Barney and the loss of the *Bonefish*.[5] She explained that the information had been pieced together from her correspondence with Lawrence, Admiral Lockwood, George Pierce, Ozzie Lynch, and several officers whom she didn't name who had firsthand knowledge of Operation Barney. Sarah knew that her letter wouldn't lessen the crushing heartbreak of the loss of those men whose lives had ended prematurely, if not needlessly. She simply wanted to share the facts she'd assembled to try to help the families understand, as she now did, what the men had faced in their hour of sacrifice.

"The loss to all of us," she wrote, "is indeed more regrettable and harder to understand since the war was all but over when the Bonefish was reported missing, for the public utterances of Adm. Nimitz say that the Japs were well defeated some weeks before the atom bomb was dropped on August 6."

In her letter Sarah shared with the families Lawrence's sentiments regarding his and his friends' belief that they had been lucky to survive the war so far and that it would be a tragedy for them not to live to see its end. Sarah also included a comprehensive overview of Operation Barney, starting with the *Wahoo*'s incursions into the Sea of Japan, which she characterized as having been turned into a highly fortified and dangerous area. Drawing on Pierce's and Lynch's letters she described the operation's tactical aspects from beginning to end, though she erred through misinformation on the two final points on which her letter ended. Nevertheless her words are a fitting conclusion to the gallant *Bonefish*'s final days in action: ". . . Adm. Lockwood sent a message to the eight subs as they went out of La Pérouse Strait, saying that disturbances in Japan Sea indicated that the Bonefish was still in the Sea* and that he had tried to send a message to her telling her how the others had left the Sea. No one went back to look for her or to see if she needed help."† To this she added, "The greatest success was that only one sub, U.S.S. Bonefish was lost. . . ."

In Tokyo, intelligence officers continued digging through records and conducting interrogations of Japanese naval personnel. They knew there had to be someone, the captain of a ship, an officer, an enlisted man, who knew something about an attack in Toyama Wan that would reveal the fate of the *Bonefish*. The team, despite working under difficult and sometimes chaotic conditions, remained optimistic that a relevant action report would eventually show up, given that action reports concerning

---

*These so-called "disturbances" are not mentioned in any of Lockwood's writings.
†George Pierce in the *Tunny* was permitted two days on standby in the Sea of Okhotsk, during which he tried unsuccessfully to raise the *Bonefish*. Lockwood apparently didn't inform Sarah of that fact.

the fates of other subs had surfaced. Adding to the difficulties, the team's efforts on behalf of ComSubPac had been harnessed to those of other Navy and Army personnel engaged in sifting material for inclusion in a massive report on Japan's technical and engineering proficiency as it related to the production of weapons and war matériel. Searching captured Japanese records for the fate of one lost submarine would prove to be a long and arduous task.

CHAPTER TWENTY

A Shining Glory

D ecember 1945 brought a change at the top of the Navy's chain of command: Admiral Nimitz replaced Admiral King as chief of naval operations. In the Pacific, Rear Admiral Allan R. McCann replaced Admiral Lockwood as ComSubPac. That McCann was a two-star rather than a three-star confirmed the diminishing role played by the Pacific sub force as the Atlantic submarine force began to meet the challenge posed by the Soviet Union.

Lockwood wasn't happy with his new assignment as the Navy's inspector general, or, as he characterized it, the Navy's top cop. For him it was the end of a life in submarines, a life he'd known for almost thirty years. He had little stomach for investigating corruption and wrongdoing in the ranks. Instead, he'd wanted a job as the Navy's submarine czar, but Nimitz wasn't interested in having one on his watch. Lockwood tried to make the best of it. He longed for his old friends and missed the camaraderie submarine service had fostered. He also missed the smells of diesel oil and the sea; he was no more fit for shore duty in postwar Washington, D.C., than he had been back in the late 1930s. Many of the skippers he had known so well, whom he had come to think of as his kin, left their wartime commands for new ones or they retired. Others went on to begin the work of building a new, modern submarine force that would need those faster, deeper-diving, quieter, more powerful submarines Lockwood had envisioned to replace those that had so ably fought in the Pacific. Also, new tactics would be needed to utilize the properties these modern subs would bring to the fleet. To man the new boats the force would need men who were proficient in electronics and advanced radars, sonars, and weapons systems. All of this lay in the future, of course. Because Lockwood knew where the force was headed, he felt trapped in his office at Main Navy; worse yet, he felt that he was being left behind. Already there was talk of an atomic-powered submarine that would revolutionize submarine warfare altogether. For men like Lockwood, those glorious days of hard-fought success against the Japanese were sliding into distant memory.

Sarah tried hard to accept that Lawrence and his men were not coming home. It was almost 1946; the war had been over for almost five months. During that time there had been no new news from the Navy Department about their fate. Captured Japanese records had not yet yielded information that would explain what had happened to the *Bonefish*. The search for POWs in Japan was coming to an end. The wait for information was agonizing.

Frustrated by the Navy's slow-moving bureaucracy, Sarah, in typical fashion, took matters into her own hands. In late December she telephoned Lieutenant General Robert L. Eichelberger, the commanding general of the U.S. Eighth Army in Yokahama, Japan, at home on Christmas leave in Asheville, North Carolina. When he answered the phone she didn't hesitate to ask for his help locating information about the *Bonefish*. He suggested that she write to him in Japan with all of the information she had accumulated so far, which he would use to conduct a search of Eighth Army records for any trace of the missing men.

In a letter to Eichelberger dated January 15, 1946, Sarah wrote, "We thought that by now some information would have come from the Japanese through claims made by their anti-submarine vessels. If the assumption that [the *Bonefish*] did not strike a mine, in which case debris would float ashore to indicate such is correct, then she was sunk by plane or ship and the Japs well know her fate. . . . These past months have indeed been of the greatest anxiety for all of us who had loved ones aboard, especially since officially we have been told only that the boys are 'missing in action' and the sub 'lost.' . . . Am I correct in having understood you to say that our forces have occupied all of Japan, which cancels our hope that the boys could yet be prisoners of war? Are our forces actually searching the many islands of the Japan Sea?"

Eichelberger replied, "I am writing . . . to inform you that to date I can only report that a check of Army and GHQ recovered personnel records and graves registration records reveals no information pertaining to Commander Edge or personnel of the submarine which your husband commanded.

"Investigations are being continued with the Naval Affairs section of General Headquarters and the Japanese Government for any possible information on this subject. . . ."

By March, further inquiries to Eichelberger proved fruitless. Undaunted and drawing on remarkable reserves of faith and strength, Sarah pressed on, more determined than ever to part what seemed like an opaque curtain of bureaucratic obfuscation surrounding the *Bonefish*'s loss. With time she'd accepted that Lawrence and his men could not have survived the sinking, and that if by some miracle some had, she knew in her heart that by now they would have been found. Lawrence was gone and nothing, not prayers, not tears, not anger, would bring him back. It was important, then, that his accomplishments were not forgotten and that his integrity and heroism be celebrated. He had placed his ideals above his fears and had died doing his duty. By taking aggressive action during his war patrols, Lawrence had personified the best of the submarine force. Like Dudley Morton, who never gave up the chase and once said, "Stay with [the enemy] till they're on the bottom," Lawrence Edge always stayed with the enemy, never mind the risks.

In Tokyo, the Navy intel teams were now working under the supervision of General MacArthur's occupation headquarters command. The team working for ComSubPac had come upon several important sources that detailed the circumstances surrounding the loss of most but not all U.S. submarines. The slog through these records, to say nothing of the interrogations of Japanese naval personnel involved in antisubmarine work, had been difficult, and pressure was building on the team to wrap up their work in Tokyo and prepare the tons of documents for shipment to Washington for further review by ONI, and for microfilming and cataloging for storage. If the team didn't find the answers they were looking for now, they might never be found after the documents arrived in Washington. Thus, it was only through their dogged determination that the documents finally yielded results.

In early May a report entitled "Tabular Summary of U.S. Submarine Losses During World War II" arrived on the desk of Admiral McCann, ComSubPac. The report gave the details and circumstances of each loss gleaned from the records and interrogations in Tokyo. While it settled the fates of most but not all of the fifty-two boats lost, the summary explained that information about the other boats wasn't available because the Japanese themselves had no direct evidence of attacks on those submarines. In those cases all ComSubPac could do was accept the fact that those boats had simply vanished.

The long nightmare of uncertainty over the *Bonefish*'s fate finally came to an end in late June 1946, more than a year after her disappearance, when a letter from the Navy arrived at Collier Road. Though Sarah had been preparing for this day, the letter's arrival struck a hard, painful note. Here at last was the end, the final, official word in cold, efficient, unadorned language.

Quoting excerpts from the official history of the Pacific Submarine Force, Admiral McCann wrote:

> *Japanese records of anti-submarine attacks mention an attack*
> *made on 18 June 1945, at 37°-18' N, 137°-25' E in Toyama*
> *Wan. A great many depth charges were thrown, and an oil pool*
> *one kilometer by ten kilometers was observed. This*
> *undoubtedly was the attack which sunk the Bonefish.*
>
> *No survivors from the Bonefish were ever found in any of*
> *the Japanese prison camps, nor did any of the repatriates from*
> *other U.S. submarine losses report seeing or hearing of survivors*
> *from the Bonefish. I cannot encourage you to believe that*
> *Commander Edge survived the loss of his ship.*

> *Please accept my deepest sympathies. Commander Edge was*
> *an outstanding naval officer and the Submarine Service has*
> *suffered a great loss.*[1]
>
> Allen R. McCann, *Rear Admiral U.S. Navy [signed]*

———

Japanese records disclosed that three sonar-equipped coast defense vessels armed with depth-charge throwers were responsible for the destruction of the *Bonefish*. According to their after-action reports, the two vessels, possibly the same two that had opened fire on the *Tunny* and the *Bonefish* near Suzu Misaki on June 17, only this time accompanied by a third unidentified ship, attacked and sank a submarine on June 18, 1945, in the area where the *Bonefish* had been patrolling in Toyama Wan.

The report stated that the three ships had been attacked with torpedoes from a submerged submarine and that the torpedoes had missed their targets. The ships then counterattacked with a savage depth charging that brought diesel oil and other debris boiling to the surface. Within hours a huge rainbow-hued oil slick could be seen from the air by planes summoned to search the area for more submarines. Unquestionably the oil had gushed from the ruptured fuel tanks of the *Bonefish*. The Japanese reported that there were no survivors. The *Bonefish* went down about eight miles from the coast in waters over a thousand feet deep. What Lawrence Edge and his crew faced in those terrible last moments, they faced with bravery and courage, giving all of their skill and strength to preserve their ship and their lives.

Before her end the *Bonefish* wrote a postscript to an outstanding career as a fighting ship. The Japanese reported on June 19 the loss of the 5,400-ton *Konzan Maru* close to where the *Bonefish* had been sunk. The reported date of the sinking of the *Konzan Maru* is incorrect, as the *Bonefish*, sunk on the eighteenth, had to have sunk the *Konzan Maru* before she herself was sunk. There is no doubt that the ship torpedoed by the *Bonefish* was the *Konzan Maru*, for no other Hellcat was in the area.

———

Though the details are sketchy it's possible to visualize the *Bonefish's* final moments.

Patrolling submerged in Toyama Wan, Edge encountered three patrol boats. He attacked, but the torpedoes missed. Alerted, the patrol boats counterattacked in force. There wasn't time to fire another torpedo salvo—the enemy's *ping, ping, pinging* sonars had the *Bonefish* trapped in a vise. *Get her down fast—four hundred feet! Rig for depth charge and silent running! Here they come!* Three sets of angry, thrashing screws swept over the descending submarine. Depth charges rained down.

Whether by luck or fate, a hull-smashing explosion closer and more powerful than any the submariners had ever experienced caused mortal damage. In the split second it took the doomed men to grasp what had happened, the sea burst into the *Bonefish* like a snarling, killing beast. *Sound the collision alarm! Blow safety! Blow bow buoyancy! BLOW EVERYTHING!* In the confusion of anger and fear, frantic efforts to avoid disaster failed. Flooded and out of control, the *Bonefish* upended. Men, tools, anything not tied down crashed into the now horizontal bulkheads at the bottom of compartments. Depth-gauge needles wound violently to their stops. The water under the sub's keel was so deep that it was beyond comprehension. Down, down she plunged until, at the limit of their endurance, her stout frames and hull, moaning and shrieking in protest, gave way to the merciless sea. Trailing skeins of air bubbles and oil, the gallant *Bonefish* with her gallant captain and crew dived into eternity.

For Sarah Edge and the grieving families a part of their lives had come to an end. Yet with these words from the mother of a *Bonefish* sailor a new one had already begun: "You and I have never met, Mrs. Edge, but we are not strangers. We share a common anguish, and a common hope, and because we are proud of our men, we bear in our hearts high courage and the knowledge that what they are and what they have done is a shining glory that will remain with us always."

AFTERWORD

hen World War II ended and the Cold War began, Operation Barney and the Hellcats faded into obscurity. In retirement and with time to reflect, Lockwood accepted that Operation Barney, though a brilliant tactical success, had not brought Japan to her knees, as he had once believed it would. And while he never doubted that the Hellcats had performed magnificently, it must surely have troubled him that the mission he had conceived and worked so hard to launch had been neither the crowning achievement of the Pacific submarine war nor free of controversy.

Lockwood, justifiably proud of his accomplishments and those of the submarine force, was sensitive to this controversy. Questions, especially those regarding what he and the Navy's senior commanders expected of Operation Barney, should be considered in light of what Admiral Nimitz and Admiral King knew about the atomic bomb. As noted earlier, King and Nimitz had been told that its use on Japan would end the war. Lockwood was not privy to this information, saying that the bomb had come as a complete surprise. Operation Barney had been planned and approved far in advance of any information about the bomb divulged to Nimitz and King. When Nimitz received his first atom bomb briefing he thought it would have little effect on the war's outcome. He was convinced that a naval blockade and conventional bombing campaign would end it. But as the war ground on and he saw how stubborn the Japanese were, he came to the conclusion that the bomb had to be dropped because it might be the only way to end the war and prevent an invasion of Japan. But because there was no guarantee that the bomb would end the war, Nimitz couldn't order Lockwood to cancel Operation Barney, for if the bomb was a dud or if it worked but didn't shock the Japanese into surrendering (which it almost didn't), or if the Soviets didn't enter the war as promised, he and King—Lockwood, too—would be faced with a Pacific campaign that would likely drag on into the fall or winter of 1945, even the spring of 1946. Thus Operation Barney had to go forward.

What, then, did Operation Barney actually accomplish? It was estimated that Japan still had some two hundred steel-hulled ships afloat in the so-called "inner zone" of empire waters. The Hellcats certainly nibbled away at that total, but it would have taken many more raids like Barney to sink all of those ships and to bring about Japan's collapse. In that light, Operation Barney seems a mere pinprick, especially compared to the atomic bomb. Yet it got off and running with an irresistible force of its own that could not easily have been terminated even if Nimitz and Lockwood had wanted to. After all, one powerful force driving the mission was the huge amount of money the FM sonar project consumed every day. As in any expensive project a usable product is needed to justify the costs. Just like the invention of the atomic bomb, the invention of FM sonar predestined its use. If Lockwood can be faulted for anything, it was his willingness to risk the lives of the Hellcat submariners to prove that an experimental sonar system he'd nurtured and helped develop could guide submarines through minefields.

With the passage of time, these issues, like Barney itself, faded into obscurity. War is war, after all, and World War II was unimaginably complicated and fought on a scale so vast that its full implications are hard to comprehend. It consumed millions of lives and billions of dollars. The commanders who planned Pacific theater strategy did so with honor and integrity. Objectivity was an essential ingredient in their planning of the multitude of operations the strategy required. The commanders had to be confident that the missions would succeed and that lives would not be squandered needlessly. When sentiment, prejudice, and emotion enter the picture they can doom a military operation to failure, to say nothing of costing lives.

Clearly, sentiment, prejudice, and emotion played a significant role in Operation Barney. Was Lockwood's need to avenge the loss of Mush Morton and the *Wahoo* the *real* driving force behind Operation Barney? In his writings after the war, Lockwood made it clear that even if he wasn't obsessed by the need for revenge it was never far from his mind and that it had had a strong influence on his decisions. I believe it's fair to say that despite the bravery and dedication of the Hellcat submariners, Operation Barney was simply not worth the risks it entailed to sink ships in the Sea of Japan and avenge Mush Morton and the *Wahoo*. Certainly it was not worth the loss of the *Bonefish*.

Nonetheless, it's important to remember that, regardless of the various reasons for Operation Barney's existence, nothing can diminish in any way the stature and integrity of Admiral Lockwood—Uncle Charlie— and his devotion to his service and his men. And nothing can diminish what the heroic Hellcats achieved during one of the most dangerous and daring operations of World War II.

Charles Lockwood addressed many of the issues surrounding Operation Barney in a book he wrote in 1951 entitled *Sink 'Em All*. A memoir of the submarine war, the book provides a broad overview of that war from Lockwood's perspective as ComSubPac. He describes Operation Barney, though not in detail, and pays a special tribute to Lawrence Edge. Four years later, Lockwood wrote the full story of Operation Barney, from its inception to its execution, in a book entitled *Hellcats of the Sea*. As he did in *Sink 'Em All*, he paid tribute to Edge, this time devoting a full chapter to the *Bonefish*. There are no revelations in *Hellcats of the Sea*, but Lockwood does admit to his desire to extact revenge on the Japanese for the loss of Morton and the *Wahoo*. He inscribed a copy of his book to Sarah Edge as follows: "To Sarah as a small tribute to the memory of a gallant submariner, Lawrence Lott Edge—God rest his soul."

Lockwood retired from the Navy in 1947. He moved to Los Gatos, California, living in a home he named "Twin Dolphins." In retirement he devoted himself to local civic issues, writing about submarines, lecturing, and hunting. He witnessed the revolution in submarine design and technology he had anticipated with the launching in 1954 of the first nuclear-powered submarine, the USS *Nautilus* (SSN-571), and, in 1960, the deployment of the first ballistic missile–firing sub, the USS *George Washington* (SSBN-598). For Lockwood, those magnificent ships were the fulfillment of what he'd envisioned back in the dark ages of submarining: true submarines endowed with unlimited submerged endurance, high underwater speed, and deep-diving capabilities, and armed with powerful torpedoes and missiles.

Lockwood died in 1967 and is buried in Golden Gate National Cemetery, San Bruno, California. Had Uncle Charlie lived to see photographs of the sunken *Wahoo* and the other recently discovered subs, he might not have believed his eyes. Then again, as an old submarine sailor, he knew that anything was possible, even that with time, the sea would give up its secrets. (Photos of the wrecks of the discovered submarines can be viewed on Google Images.)

Lockwood's right-hand man, the brilliant Richard G. Voge, after serving as operations officer of ComSubPac, joined the staff of CNO Nimitz. Voge took on another of Lockwood's pet projects, an official administrative and operational history of the submarine force. After completing this massive work, Voge retired from the Navy in 1946 with the rank of rear admiral. He died in 1948. The histories Voge compiled were never published as originally planned but instead became the massive volume entitled *United States Submarine Operations in World War II*, published in 1949 by the United States Naval Institute.

Chester W. Nimitz, the Navy's last fleet admiral, retired in 1947. He died in 1966 and, like Lockwood, is buried in Golden Gate National Cemetery, San Bruno, California. The Navy's current fleet of *Nimitz*-class nuclear-powered aircraft carriers are named in his honor.

Fleet Admiral Ernest J. King retired in 1945 after relinquishing his post as CNO to Admiral Nimitz. King died in Kittery, Maine, in 1956.

Captain Earl T. Hydeman, commander of Hydeman's Hellcats, died in 1993. He was buried with full military honors in Arlington National Cemetery.

Sarah Edge, accompanied by her daughter, Sarah (Boo), and her son, Lawrence Jr., sponsored the *Bonefish II* (SS-582), launched on November 22, 1958, at New York Shipbuilding Corporation, Camden, New Jersey. The *Bonefish II*, a diesel-electric-powered *Barbel*-class submarine, was one of the last of its type in the U.S. Navy. The Navy decommissioned her in 1988.

Sarah Edge maintained an active life in Atlanta society and wrote a biography of her grandfather, entrepreneur and developer Joel Hurt. Years later she remarried and enjoyed her six grandchildren, who now have children of their own. Sarah died in 1985.

Most, but not all, of the Hellcat submariners are dead now. Operation Barney itself, the most daring submarine raid of all time, is largely forgotten. Though the *Bonefish* and her crew rest in the deep waters of the Sea of Japan, they and the men of the *Sea Dog, Crevalle, Spadefish, Tunny, Skate, Flying Fish, Bowfin,* and *Tinosa* are toasted, as are all the subs and submariners of World War II, but especially those on eternal patrol, whenever and wherever submarine veterans gather to remember what their comrades did when they were Hellcats.

---

*The *Bowfin* is one of only a few fleet-type submarines saved from the breakers' yard and placed in service as a museum ship. She can be visited at the USS *Bowfin* Submarine Museum and Park, Pearl Harbor, Hawaii.

APPENDIX ONE

Japanese Naval and Merchant Vessels Sunk by the USS *Bonefish* as Compiled by JANAC

Comdr. T. W. Hogan

Date	Name	Type	Tonnage
September 27, 1943	*Kashima Maru*	transport	9,908
October 10, 1943	*Teibi Maru*	transport	10,086
October 10, 1943	*Isuzugawa*	cargo	4,212
November 29, 1943	*Suez Maru*	cargo	4,646
December 1, 1943	*Nichiryo Maru*	passenger-cargo	2,721
April 26, 1944	*Tokiwa Maru*	passenger-cargo	806
May 14, 1944	*Inazuma*	destroyer	1,950

Comdr. L. L. Edge

Date	Name	Type	Tonnage
July 30, 1944	*Kokuyo Maru*	tanker	10,026
September 28, 1944	*Anjo Maru*	tanker	2,068
October 14, 1944	*Fushimi Maru*	cargo	2,542
June 13, 1945	*Oshikayama Maru*	cargo	6,892 (Sea of Japan)
June 18, 1945	*Konzan Maru*	passenger-cargo	5,488 (Sea of Japan)

Total   12 vessels   61,345 total

APPENDIX TWO

Japanese Naval and Merchant Vessels Sunk by the Nine Hellcat Submarines in the Sea of Japan as Compiled by JANAC (includes those sunk by the *Bonefish*)

Submarine	Date	Name	Type	Tonnage
Sea Dog				
	June 9	*Sagawa Maru*	cargo	1,186
	June 9	*Shoyo Maru*	cargo	2,211
	June 11	*Kofuku Maru*	cargo	753
	June 12	*Shinsen Maru*	cargo	880
	June 15	*Koan Maru*	cargo	884
	June 19	*Kokai Maru*	cargo	1,272
Crevalle				
	June 9	*Hokuto Maru*	cargo	2,215
	June 10	*Daiki Maru*	cargo	2,217
	June 11	*Hakusan Maru*	cargo	2,211
Spadefish				
	June 10	*Daigen Maru No. 2*	cargo	4,273
	June 10	*Unkai Maru No. 8*	cargo	1,293
	June 10	*Jintsu Maru*	cargo	994
	June 14	*Seizan Maru*	cargo	2,018
	June 17	*Eijo Maru*	cargo	2,274

continued...

Submarine	Date	Name	Type	Tonnage
Skate				
	June 10	*I-122*	submarine	1,142
	June 12	*Yozan Maru*	cargo	1,227
	June 12	*Kenjo Maru*	cargo	3,142
	June 12	*Zuiko Maru*	cargo	887
Bonefish				
	June 13	*Oshikayama Maru*	cargo	6,892
	June 18	*Konzan Maru*	cargo	5,488
Flying Fish				
	June 10	*Taga Maru*	cargo	2,220
	June 11	*Meisei Maru*	cargo	1,893
Bowfin				
	June 11	*Shinyo Maru No. 3*	cargo	1,898
	June 13	*Akiura Maru*	cargo	887
Tinosa				
	June 9	*Maru*	cargo	2,211
	June 12	*Maru*	cargo	880
	June 20	*Kaisei Maru*	cargo	884
	June 20	*Taito Maru*	cargo	2,726

Total   28 vessels   57,058 total

No totals are shown for the *Tunny*, as she sank no ships. The *Spadefish* sank the Russian *Transbalt*, an 11,400-ton cargo ship. The Hellcats also sank more than a dozen small craft by gunfire.

APPENDIX THREE

Full Text of the Letter from Sarah Simms Edge to the Families of the Crew of the USS *Bonefish*

[Undated]

The information relative to the loss of the USS Bonefish, which I have been able to obtain, may be of interest to you since so many of you have requested the details which I have learned. It is pieced together from letters from Vice Admiral Chas. A. Lockwood, Jr., Commander Lawrence L. Edge, captain of the USS Bonefish, Commander Geo. Pierce, captain of the USS Tunny, and others who were on the raid in the Japan Sea in June of 1945.

The loss to all of us is indeed more regrettable and harder to understand since the war was all but over when the Bonefish was reported missing, for the public utterances of Adm. Nimitz say that the Japs were well defeated some weeks before the atom bomb was dropped on August 6. Comdr. Edge wrote in April while on patrol, "There is the feeling among all of us that we have been lucky enough to survive the war so far; it would be such a shame not to last for the remainder and thus live through the whole thing."

The Japan Sea had been secured (closed) to our subs for the past two years, after the loss of the USS Wahoo there, because it was decided by the Navy that the risks involved were too great for the gains received. It was known that during those two years the Japs had made their Sea a highly fortified and more dangerous area. Submarine high command had wished for some time to send our subs back into that Sea, so a gear to contribute to this safety of our submarines was developed and pushed. From the fall of 1944 to May 1945, ten subs were equipped with this gear after their overhauls. These were the *Skate, Tunny, Crevalle, Bowfin, Flying Fish, Tinosa, Sea Dog, Spadefish, Bonefish,* and *Seahorse.*

The *Bonefish* received her gear in Pearl Harbor* in March after she had completed her overhaul in San Francisco. From Pearl Harbor she went to Guam, where Adm. Lockwood made a couple of short practice runs on her to observe this equipment. In April he sent the *Bonefish* and the *Seahorse* to clear [plot] a path through Broughton [Tsushima] Strait, which is 65 miles long, so that the subs could enter the Japan Sea with relative safety as far as mines were concerned. The *Seahorse* was so badly damaged by enemy escort vessels, that she was forced to return to base and was unable to join the Japan Sea raid. The *Bonefish,* however, was able to complete that part of her "special mission" in April with much glory and the Navy Cross was won for her work, though none aboard knew of their award when they left for their June raid.

I say "that part of her special mission," because after the *Bonefish* completed the work of [charting minefields], she and the other subs with this equipment which were then out on patrol were radioed to return to port earlier than their orders originally stated. The rest periods for those of the nine subs that were on patrol were cut short of their normal two weeks. Some had only six days. "The *Bonefish* had a short rest period (8 days) because she had to make the schedule with the rest of us." Comdr. Edge was much disappointed that his crew would not have their usual and much needed two weeks' rest after their "most dangerous and exhausting patrol," during which

*The *Bonefish* was actually outfitted with FMS in San Francisco.

*he was continually concerned for the safety of the boat
and crew.*

*The nine boats made a few practice runs and started out,
some felt before they were ready. The Bonefish left Guam on
May 28. The boats were divided into three groups of three
each. The Bonefish was in the second group headed by the
Tunny. The groups went through the strait at the rate of one a
day. The Tunny's group entered, therefore, on the second day,
June 13. No subs [in that group] opened fire on Jap shipping
until June 9 when all boats were in their respective areas,
different bays on the west coast of Honshu.*

*Soon after the operations had begun the Japs radioed to
their ships and planes of our activity in the sea. Our subs
received this message almost as quickly as did the Japs. It was
during this raid that the Japs announced the sinking of six or
seven subs.*

*The second group headed by the Tunny held a rendezvous
out in the Japan Sea on June 16th, patrolled together on the
17th. On the 18th, they met again 50 miles northwest of Noto
Peninsula. On this night, another sub reported to the "pack
commander," Geo. Pierce of the Tunny, that she had seen
enemy ships entering Toyama Wan. "The Bonefish went in
after them because that was her assigned patrol area." Another
sub was assigned half of this bay area for two days only and was
nearby until she had fired her last torpedo. At this time the
Bonefish was known to be "afloat."*

*"Toyama Wan was no more hazardous than any other place
except it was a bay and as such had only one direction to go to
get out." The depth of this bay is 350 fathoms.*

*"No one knows just what happened to the Bonefish. We
thought that some information might come from the Japanese
after the war was over through claims made by their antisubmarine vessels. There is one third of a chance that they
struck a mine. There is no way to tell just where the boat might
have been sunk."*

On June 18, four days before the Bonefish was assumed missing, she reported having sunk a large cargo vessel and a transport. "She could have easily sunk more."

The nine subs were scheduled to hold a rendezvous on or about the night of June 21 to decide their exit through La Pérouse Strait, whether surfaced or submerged. The Bonefish did not appear, but it seems the foggy weather which existed made the other subs think nothing of her absence. The Tunny and the other subs went on out of the Japan Sea and then, seeing the continued absence of the Bonefish, Comdr. Pierce waited there for two days and tried to contact her by radio. I have heard that the Bonefish had radio and radar trouble all during the patrol. Commander Submarine Pacific Fleet, Adm. Lockwood, sent a message to the eight subs as they went out of La Pérouse Strait, saying that disturbances in Japan Sea indicated that the Bonefish was still in the Sea and that he had tried to send a message to her telling how the others had left the sea. No one went back to look for her or to see if she needed help.

All nine subs received the Navy Cross for this raid, including the Bonefish which probably sunk several ships besides the two which she reported at the rendezvous on June 18th. The Tunny, leader of group two, was the only one of the nine to sink not one enemy ship. A total of [twenty-eight] ships were destroyed.

A conference of all officers from the eight returning boats was held with Adm. Lockwood in [Pearl Harbor]. Success within the sea was reported as far as damage done to the enemy was concerned. As for the [FMS] gear, it was not infallible. The greatest success was that only one sub, USS Bonefish, was lost, whereas I am told that "at least" half of those entering the raid were not expected to return. Several have written me or told me the equivalent of the following quotation from one comdr.: "You are correct about the patrol being dangerous. I felt that I would not get back. I did not tell anyone about it but it is the case."

Comdr. Pierce wrote me that the "whole affair was on a

volunteer basis." In other more accurate words, the subs with
the above equipment were called into port to participate and
none dared refuse. As for the Bonefish, Comdr. Edge was told
by the Submarine Force Personnel Officer the day he arrived in
Pearl Harbor in March that he would be sent to a staff job as
electronics officer immediately after his April patrol. Despite
this, on reaching Guam before that patrol began, Adm.
Lockwood told him personally that he would have to take the
Bonefish on this one more patrol, and ordered her on the raid.
This is the "volunteer basis" on which Comdr. Edge took the
Bonefish on the raid in the Japan Sea. He wrote me of his
disappointment on learning that he had another patrol yet
to make.

From another source: "We were ordered to make the June
raid the same way we were ordered to make any other patrol
except we had all this special charts and gear to make the run. I
was not asked particularly whether I wanted to make the run or
not. It has always been the policy for any skipper who wanted to
give up command to merely say so and he would be relieved.
However, I was not asked directly whether I wanted to make the
patrol."

The USS Bonefish made a brilliant record on her eight war
patrols deep in the enemy's territory. Her two commanders had
won for her six Navy Cross medals, a Silver Star, a Bronze Star,
and a Navy Unit [Commendation]. For Comdr. Edge's first
patrol on her, June 25 to August 13, 1944, the Bronze Star was
awarded for the sinking of enemy ships totalling 20,000 tons,
and for damaging 7,500 tons. For his second patrol which took
place in the Sulu Sea and around the Philippines, September 15
to October 27, 1944, his first Navy Cross was awarded for the
sinking of enemy ships totalling 22,000 tons, and for damaging
additional vessels totalling 7,000 tons. The danger involved in
the areas of this patrol and the fact that two of our own fliers
were rescued by the Bonefish were considered in this award.

It was during this patrol that General MacArthur first

landed in the Philippines on Leyte Island. Comdr. Edge's third and fourth patrols in the Bonefish are discussed above and each won additional Navy Cross medals. According to Adm. Lockwood, the Bonefish, under the command of Comdr. Edge, is known to have sunk four tankers, five freighters, two transports, and five small craft, also with the damaging of a minelayer and a large tanker.

Sarah Simms Edge [unsigned]

APPENDIX FOUR

An Edited Address
by Fleet Admiral Chester W. Nimitz
Delivered on Navy Day, October 27, 1945

Almost four years ago, when they struck a treacherous blow at our sea power at Pearl Harbor, the Japanese galvanized America into action. From their homes in every quarter of the United States men and women hastened to volunteer in the Armed Forces of our country. Bravely they went forth on our ships, they landed on hostile shores, and fought with the knowledge that it might—and in some cases, it would—cost them their lives.

On Navy Day this year—let us remember those officers and men whose spirits look down on us from some Valhalla of fighting men. They gave their utmost—they gave their all. The impressive record of the ships they manned is long. You have heard the names of some of them. There are others—now lost—which will be long honored in our history.

The battleship *Oklahoma*, the cruiser *Houston*, the carrier *Lexington*, the escort carrier *Lipscomb Bay*, the destroyer *Laffey*, the destroyer escort *Samuel B. Roberts*, the minesweeper *Emmons*, the submarine *Bonefish* which, in the closing days of the war, entered the Sea of Japan through dangerous minefields to cut the last links between Japan and her resources on the Asiatic mainland.

These ships, and others, went down in the far places of the earth, each of them fighting in the best traditions of our Navy.

I cannot name them all. But, those that have been named can stand for all the others on which men have died at their battle stations. Those men died that we might have peace and security. They helped to shape in the forges of war the great sea power which is ours today.

C. W. Nimitz

APPENDIX FIVE

Award and Commendation to Commander Lawrence L. Edge by Vice Admiral Charles A. Lockwood Jr.

The Commander Submarine Force, Pacific Fleet, has the honor to award the Submarine Combat Insignia and to commend in absentia

COMMANDER LAWRENCE L. EDGE, U.S. NAVY,

FOR SERVICES AS SET FORTH IN THE FOLLOWING CITATION:

"The U.S.S. BONEFISH failed to return as scheduled from an offensive war patrol which she was conducting in restricted waters, heavily patrolled by enemy air and surface forces. This vessel has continuously distinguished herself since her first appearance in enemy water by her successful and relentless attacks against the enemy, and it is definitely known that the BONEFISH was pursuing just such bold and aggressive tactics up until the time she was declared missing.

"As commanding officer of the U.S.S. BONEFISH, Commander Lawrence L. Edge's skill, daring and wholehearted devotion to his service and to his country contributed directly to the vessel's many successful attacks against the enemy. The Commander Submarine Force, Pacific Fleet, forwards this commendation in recognition of his splendid leadership and courageous performance of duty, which were in keeping with the highest traditions of the Naval Service."

C. A. Lockwood, Jr., Vice Admiral, U.S. Navy. [signed]

ACKNOWLEDGMENTS

Many individuals played an important role in the writing of *Hellcats*. Most important of all are Lawrence Lott Edge Jr. and Sarah Edge Shuler, son and daughter of Lawrence Lott Edge and Sarah Simms Edge. Larry, Sarah, and their spouses opened their hearts and their homes to me, welcomed me as a member of the Edge family, and provided every kindness and assistance I needed to write *Hellcats*. Larry and Sarah's mother had kept a voluminous cache of correspondence and documents from the war years, which I was privileged to read and study at length. I was staggered by the sheer amount of material in this cache, all of it in near pristine condition after having survived numerous family moves and storage in less than ideal conditions over a span of more than sixty-five years. Perusing this material I had the uncanny feeling that Sarah Edge had carefully saved it not just because it was a touchstone to her former life, long vanished, but also because she may have believed that someday someone would want to write a book about what Lawrence and his shipmates accomplished.

To be sure, *Hellcats* is not the first book ever written about Operation Barney. Charles A. Lockwood's pioneering *Hellcats of the Sea* was published in 1955 and later made into a Hollywood movie, entitled *Hellcats of the Navy*. The film starred Ronald Reagan and Nancy Davis (later, Nancy Reagan) as the love interest. Though Lockwood's book has long been out of print, his telling of the story is an essential road map to the navigation of Operation Barney's complex, multilayered history. The Edge family archive provided another kind of road map: a vast and rich trove of personal and poignant wartime correspondence between Lawrence and Sarah overflowing with a man's love for his wife and children. They also provided an intimate view of the emotions experienced by a wartime sub skipper that I believe are unique to the submarine service. In addition, the letters and official documents in the collection reveal the inner workings of a cold Navy bureaucracy at war with the Japanese and sometimes with itself. The exception to this is the warm correspondence between Sarah Edge and Charles Lockwood. All of this material gave the already compelling story of Operation Barney a depth and texture it would not otherwise have had. For all of the help and kindnesses shown me by the Edge family, I am deeply grateful.

I am also deeply grateful to Dr. Kurt S. Maier, Senior Cataloger at the Library of Congress in Washington, D.C. He provided copies of documents relative to Operation Barney from the papers of Charles A. Lockwood archived at the library. Dr. Maier was tireless in his searches and findings, and provided valuable assistance without which *Hellcats* could not have been written.

I also want to thank Barbara Hydeman Barnes, Patricia Hydeman Barry, and her husband, Bob, for providing personal documents and photographs related to the career of Barbara and Patricia's father, Captain Earl T. Hydeman, and his participation in Operation Barney. Their deep interest and gracious response to my requests are greatly appreciated. Also, I want to thank Patricia Fornshell, whose father, Allen George Maghan, F1c, perished on the *Bonefish*. Patricia provided family documents that in the early planning of *Hellcats* offered insights into the missing-in-action stage of the *Bonefish*'s loss.

Thanks are also due to William E. Scofield, who served in the *Bonefish* under her first skipper, Commander T. W. Hogan. Bill set me on the right course to locate material on the ship's early career and *Bonefish* family members. Also thanks to John Clear, EMC (SS), U.S. Navy (retired), for his generous assistance with the World War II submarine patrol reports for the Hellcats and other boats. Author Steven Trent Smith provided valuable documents related to Operation Barney from his private collection. Denise Clark and Michele Mazanec at the United States Naval Academy Alumni Association located biographical materials on various academy graduates who skippered submarines during World War II. Thanks, too, to the staff of the Nimitz Library at the United States Naval Academy for their generous loan of rare microfiche of ComSubPac's administrative history; and to the staff of the Scripps Institution of Oceanography Library, University of California, San Diego, La Jolla, California, for providing copies of documents originally published by the University of California Division of War Research (UCDWR), San Diego, California.

Deep appreciation goes to my literary agent, Ethan Ellenberg, for his advice and counsel. And to Brent Howard and the editorial and production staff at Penguin/NAL for their hard work and expert guidance in the publication of *Hellcats*.

As always, I thank my wife, Karen, for her abiding interest in and help with a multitude of issues both big and small as this project first developed from an idea, then stretched out over the course of many months. Finally, a special thanks to my dad, a submariner's submariner, who, during World War II, served in the USS *Rasher*.

BIBLIOGRAPHY

Books

Alden, John D. *The Fleet Submarine in the U.S. Navy: A Design and Construction History.* Annapolis, Maryland: Naval Institute Press, 1979. (The only book of its kind that presents in arresting detail the development of the U.S. fleet-type submarine and why they looked and performed like they did. A fascinating study.)

Allen, Thomas B., and Norman Polmar. *Code-Name Downfall: The Secret Plan to Invade Japan—and Why Truman Dropped the Bomb.* New York: Simon and Schuster, 1995. (A concise but detailed account of the planning for the invasion of Japan, with statistics of both offensive and defensive forces including estimated casualties.)

Beach, Edward L. *Submarine!* New York: Henry Holt, 1946. (One of the best books ever written about World War II U.S. submarine operations, by a man who lived it.)

Blair, Clay, Jr. *Silent Victory: The U.S. Submarine War Against Japan.* Philadelphia and New York: Lippincott, 1975. (The definitive history of the submarine war. Blair also does a good job of making sense of the ever-changing command structure, which influenced operations for both good and bad.)

Brackman, Arnold C. *The Other Nuremberg: The Untold Story of the Tokyo War Crimes Trials.* New York: Quill/William Morrow, 1987. (Brackman revisits the issues that made these trials so controversial in their time.)

Costello, John. *The Pacific War: 1941–1945.* New York: Quill, 1982.

Cragon, Harvey G. *The Fleet Submarine Torpedo Data Computer.* Dallas, Texas: Cragon Books, 2007. (An in-depth tour of and working guide to the type of TDCs installed in U.S. subs. Not for the casual reader, as expertise in trigonometry and electrical theory are essential to a complete understanding of this complicated analog computing machine.)

Dull, Paul S. *A Battle History of the Imperial Japanese Navy (1941–1945)*. Annapolis, Maryland: Naval Institute Press, 1978.

Friedman, Norman. *U.S. Submarines Through 1945: An Illustrated Design History*. Annapolis, Maryland: Naval Institute Press, 1995.

Hastings, Max. *Retribution: The Battle for Japan, 1944–45*. New York: Vintage, 2009.

Holmes, W. J. *Double-Edged Secrets: U.S. Naval Intelligence Operations in the Pacific During World War II*. Annapolis, Maryland: Naval Institute Press, 1979.

Lockwood, Charles A. *Down to the Sea in Subs: My Life in the U.S. Navy*. New York: W. W. Norton and Company, 1967.

————. *Sink 'Em All: Submarine War in the Pacific*. New York: Dutton, 1951.

————, and Hans Christian Adamson. *Hellcats of the Sea*. New York: Greenberg, 1955. (Lockwood's personal account of Operation Barney: the road map through it.)

Moore, Stephen L. *Spadefish: On Patrol with a Top-Scoring World War II Submarine*. Dallas, Texas: Atriad Press, 2006.

O'Kane, Richard H. *Wahoo: The Patrols of America's Most Famous Submarine*. Novato, California: Presidio Press, 1987. (O'Kane's experiences as Dudley Morton's exec. Provides keen insights into Morton's thinking and doing.)

Parillo, Mark P. *The Japanese Merchant Marine in World War II*. Annapolis, Maryland: Naval Institute Press, 1993. (A very good study of Japan's wartime merchant marine operations, including tables and statistics that provide a clear picture of the merchant marine's successes and failures.)

Potter, E. B. *Nimitz*. Annapolis, Maryland: Naval Institute Press, 1976. (The definitive biography of CinCPac; with valuable insights into the thinking and political maneuverings of the Navy's high command.)

Rhodes, Richard. *The Making of the Atomic Bomb*. New York: Simon and Schuster, 1986.

Roscoe, Theodore. *United States Submarine Operations in World War II*. Annapolis, Maryland: Naval Institute, 1949. (A dated but still valuable source for information about the U.S. sub war in all its facets—especially valuable for JANAC sinking statistics and much more.)

Sasgen, Peter T. *Red Scorpion: The War Patrols of the USS Rasher*. Annapolis, Maryland: Naval Institute Press, 1995.

Scofield, William F. *Dear Ma!* Facet History Publications, 2007. (A personal memoir of the author's service aboard the USS *Bonefish*, and "dedicated to life before television"; a lively commentary on the submarine service.)

Shirer, William L. *End of a Berlin Diary.* New York: Popular Library, 1961.

Smith, Steven Trent. *Wolf Pack: The American Submarine Strategy That Helped
Defeat Japan.* Hoboken, New Jersey: John Wiley and Sons, Inc., 2003. (Smith's
book gives a detailed account of U.S. submarine wolf-packing, and also the
history of and the individuals responsible for the development of FM sonar.)

Winton, John. *Ultra in the Pacific: How Breaking the Japanese Codes and Ciphers
Affected Naval Operations Against Japan.* Annapolis, Maryland: Naval Institute Press, 1993.

Archives, Private and Public

The correspondence of Lawrence L. Edge and Sarah S. Edge. Letters from the collection of Lawrence L. Edge Jr. and Sarah Edge Shuler.

The correspondence of Sarah S. Edge and Charles A. Lockwood Jr., other naval
officers, and U.S. Navy bureaus.

The correspondence of Sarah S. Edge and various families of the crew of the USS
Bonefish.

The service records of Lawrence Lott Edge, courtesy of Lawrence L. Edge Jr.

The papers of Charles A. Lockwood. The Manuscript Division of the Library of
Congress, Washington, D.C. (Lockwood's day-to-day triumphs and travails
described in correspondence with his colleagues both in the United States and
at various bases throughout the Pacific theater of operations.)

Hyper War on the Web at www.ibiblio.org. "United States Strategic Bombing Survey, Naval Analysis Division, Summary Report, July 1, 1946." And "Reports on
the U.S. Naval Technical Mission to Japan." And "Japanese and Merchant Shipping Losses During World War II by All Causes."

National Archives II, Modern Military Records, College Park, Maryland. Record
Group 313: ComSubPac operation orders related to Operation Barney. (Covers
dates, timing, codes, and contingencies for the execution of Barney.)

Operational Archives Branch, Naval Historical Center, Washington, D.C. "Current
Doctrine, Submarines," prepared by Commander Submarine Force, Pacific
Fleet, February 1944. (An excellent source for information on World War II
shipboard submarine operations.)

Scripps Institution of Oceanography Archives, University of California, San Diego,
La Jolla, California: Office of Scientific Research and Development, National

Defense Research Committee. "FM Sonar," by Malcolm C. Henderson and Charles A. Hisserich. Abstracts, operating and repair manuals for submarine-installed FM sonar. Published by the University of California Division of War Research at the U.S. Navy Radio and Sound Laboratory, San Diego, California, September 4, 1943. (A fascinating look at the technical developments leading to the development of FM sonar and its operation and maintenance aboard ship.)

Miscellaneous

Central Intelligence Agency. "The Final Months of the War with Japan." See Center for the Study of Intelligence at www.cia.gov. (An in-depth report on the circumstances leading to Japan's surrender and the internal decision-making process behind President Truman's decision to use the atomic bomb.)

Submarine patrol reports on CD. A collection of more than 63,000 pages of submarine patrol reports, compiled from original U.S. government microfilms by John Clear, EMC (SS), U.S. Navy (ret.). (An invaluable resource for any submarine historian—fascinating reading in their own right.)

United States Naval Academy *Lucky Bag* (yearbook) biographies for Lawrence L. Edge and Dudley W. Morton.

Vintage Newspaper Collections

Various writers and articles. *Atlanta Georgian*, June 15, 1938; *Atlanta Journal*, April 24, 1938, and June 16, 1938; *Columbus Inquirer* (Columbus, Georgia), June 29, 1938; *Atlanta Constitution* (Atlanta), June 16, 1938; July 25, 1938; *Sunday Ledger* (Columbus, Georgia), April 24, 1938.

Newspaper and Journal Articles

Ambrose, Stephen E. "The Bomb: It Was Death or More Death." *New York Times*, August 5, 1995.

"Hiroshima, 50 Years Later." *New York Times* editorial, August 6, 1995.

Palmer, Kyle. "Sub Flotilla Returns After Taking Nip Toll." *Los Angeles Times*, August 12, 1945.

Weber, Mark. "Was Hiroshima Necessary?" *Journal of Historical Review*, May–June 1997 (Vol. 16, No. 3). Institute for Historical Review.

OFFICIAL SAILING LISTS
OF THE HELLCAT SUBMARINES

HYDEMAN'S HEPCATS

SEA DOG

Argo, Wesley B., Lt.(jg)

Bass, Joseph W., StM1

Bishop, Richard O., SC1

Brown, William S., Lt.

Bryant, Douglas A., MoMM3

Buesing, Von Clarence, MoMM3

Carter, Robert L., Jr., F1c

Cogden, Anthony E., F1c

Crowley, Harold F., PhoM2

Dell, Andrew F., CTM

Duckworth, Edward W., Lt.(jg)

Dunham, Paul P., S1c

Eilert, Morgan B., TME2

Fails, Earl F., F1c

Ficket, Albert W., Jr., RT2

Fisher, Ronald G., TM2

Fitzpatrick, Thomas F., GM2

Fontnote, Sidney T., MoMM1

Gibson, Kenneth L., TM2

Glass, Orval A., S1c

Gressman, George A., MoMM1

Griffith, Eddie E., RM2

Hammel, Paul A., S1c

Harrell, Max A., FCS2

Harry, Frank J., EM2

Heebner, Newell, G., CQM

Heiden, Walter G. E., CEM

Heller, David A., S1c

Hessman, Robert D., MoMM1

Hinchley, John F., Lt.(jg)

Hindert, Edward M., Lt.

Hinkel, Russell D., EM3

Hoyt, William D., RM2

Hydeman, Earl T., Cdr.

Ignaisak, Bernard A., TM3

Jacek, Martin P., MoMM3

Johnson, Richard E., F1

Jones, Oliver H., PhM1

Juniper, Albert J., Y1

Karn, Fred W. III, CMoMM

Kiesel, Earl J., MoMM3

Kornichuk, Arthur, WOE

Kral, Elmer S., MoMM1

La Bore, Louis J., F1c

Lanczky, William A., SC3

Lennox, William, MoMM2

Lewis, Lester B., CTM

Lupe, Theodore, Jr., S1c

Lynch, James P., Lt.

McAuliffe, William J., F1c

McCormick, Paul L., F1c

McKenzie, Thomas S., Jr., Y3

McLarty, Pat, TM2

Meacham, Arthur J., MoMM1

Misch, Frank E., RT3

Moser, Ben, SM2

Mracek, Edward, MoMM1

Murzic, William R., RM3

Nicholson, William H., RM3

Nicodemus, Ivan, Jr., TM3

Noble, Willie Z., F1c

Pagnam, James R., F1c

Parker, Earl W., MoMM1

Peterson, Leonard "J," QM3

Powell, Cecil O., TM1

Prince, James B., EM2

Ptaszynski, Arthur C., S1c

Reed, Kelly B., Lt.

Ripple, Robert E., F1c

Roberts, William M., EM3

Rue, Dwayne T., EM1

Rutledge, Charles W., S1c

Saunders, Charles M., Jr., S1c

Sawyer, Albert J., RT2

Schleuter, Vernon E., F1c

Schwind, William F., Ens.

Shelby, Claude L., SC3

Sims, James A., RM1

Steppe, Raymond, MoMM2

Swain, Robert R., EM3

Thomas, Louis E., EM1

Tompkins, Edward E., GM3

Truscelli, Anthony F., Jr., TME3

Voss, William A., S1c

Williams, Walter W., GM3

Wilson, Harold L., QM2

Wilson, Richard D., StM1

Zimmerman, Florin W., EM2

CREVALLE

Adams, William, L., S1c

Barnes, Frank, MoMM1

Bessette, Roland P. P., TM3

Biehl, Henry T., CRT

Bolin, Willis G., TM3

Bowe, R. E., Lt.(jg)

Brooks, Marvin M., GM1

Brophy, John P., F1c

Brown, Robert J., S1c

Coyer, James W., S1c

Flaherty, Joseph E., CMoMM

Fletcher, Chester J., S1c

Folse, John S., RM3

Freeman, Edgar A., TM2

Fritchen, William L., GM2

Gaines, Robert E., MoMM2

Gogul, Frank S., MoMM1

Goodman, Francis W., S1c

Graham, Ivan H., MoMM3

Helix, Max R., MoMM1

Hildebrand, Charles F., EM2

Howard, Stephen A., FC3

Howie, Robert C., MoMM2

Jaycox, John A., StM2

Jenigen, Albert, F1c

Jones, Jerome L., S1c

Katchis, Jim "A," QM3

Keane, Edward F., S1c

Kneisly, George E., S1c

Langfieldt, Maurice E., TM3

Larsen, James L., MoMM2

Lenatz, John J., TM2

Lord, E. R., Lt.(jg)

Loveland, R. A., Lt.(jg)

Lubinsky, Walter, EM3c

Malin, Ralph, F1c

Mazzone, W. F., Lt.(jg)

McGowan, Thomas F., Jr., TM1

McHugh, John, F1c

McNorgan, Joseph W., EM3

Minaker, Russel S., RT1

Minor, Bert E., StM2

Morin, G. F., Lt.

Mushett, Robert W., Y1

Newell, Richard P., S2

O'Brien, Joseph F., CSM

Osborne, Cedric H., CPhM

Pablo, Marcelo A., SC1

Plachowicz, Frank A., GM2

Polk, Lloyd E., RM3

Raider, A. J., Lt.

Rennecke, Wyman A., EM1

Reynolds, Rodney R., SM3

Roarback, Gilbert L., TM2

Schaeffer, John W., III, MoMM1

Schwarz, Robert F., EM3

Scisco, Clayton S., MoMM3

Secl, J., Lt.(jg)

Seymour, Jack M., Cdr.

Sherick, Albert M., EM3

Silvia, Richard G., TM2

Sinclair, Joe M., Jr., EM3

Singer, Jack W., EM3

Slyter, Gilbert G., EM3

Smith, John V., S1c

Stagman, Paul L., EM2

Starnes, Kenneth J., MoMM3

Steinmetz, E. H., Cdr.

Stemler, Milton D., RT3

Stokes, Frank H., SC2

Stutzman, Gerald W., RM1

Thomas, Everett A., QM1

Thompson, Robert, Jr., F2c

Thompson, William H., Bkr3

Tomlin, George L., EM2

Truman, Horace L., MM1

Wagenbrenner, Fred, EM2

Weber, Russel F., F1c

Westbrook, E. M., Lt. Cdr.

Westerlund, Alfred, MoMM3

Wheelus, Roy C., QM2

Wiesniewski, Francis W., MoMM2

Williams, George F., CEM

Woodhouse, Robert R., Y3

Zessman, Sam, TM3

SPADEFISH

Armstrong, Edward R., PhoM2

Asher, Warren J., EM3

Babb, Maurice L., Jr., F1c

Barton, Thad R., TM3

Bassett, Richard H., EM3

Brennies, Harry J., MoMM3

Brewer, John B., Jr., MoMM3

Brooks, Sie, Jr., StM1

Buncke, Harry J., Ens.

Bynum, William T., StM1

Carney, Hugh P., S1c

Case, Joseph B., RT1

Casey, James D., EM3

Charles, Walter J., Jr., FC3

Cole, James D., F1c

Cruze, Herman F., Jr., EM2

Cunningham, Edwin W., MoMM2

Cuthbertson, John M., TM2

Decker, Daniel D., Lt.

Dependahl, Leonard E., EM1

Dix, Ramond E., Ens.

Dunleavy, Anthony, Jr., MoMM3

Eimermann, Willard C., CBM

Falconer, LeRoy D., WOM

Fellows, Richard D., Lt.

Fletcher, James W., QM2

Gamby, Orville R., MoMM2

Germershausen, William J., Jr., Cdr.

Gouker, Zelbert, SM2

Graf, Charles A., S1c

Griffith, Charles C., CMoMM

Harbison, Joseph A., MoMM2

Holeman, Victor R., MoMM1

Hord, Cleveland M., F1c

Ingberg, Norval O., S1c

Ives, Victor L., CPhM

Keeney, William J., Jr., RT2

Kite, Vernon J., TM1

Kreher, Emery A., TM1

Kreinbring, Irwin H., Y1

Lacroix, Edward J., Lt.(jg)

La Fose, Murphy, F2c

La Rocca, Albert G., EM3

Lester, Clifford R., MoMM3

Lewis, Edgar L., CGM

Lundquist, Hugo C., TM2

Majoue, Paul H., Jr., CRM

Massar, Bernard A., GM2

McMahon, Wallace F., TM3

Melstrand, Howard W., S1c

Mikesell, Robert E., S2c

Miller, Thomas H., MoMM2

Moody, Roy H., EM1

Morrison, James Walter, S1

Mullen, Wallace F., TM3

Nesnec, John, TM2

Noonan, Maurice A., GM3

Olah, Andrew, FC1

O'Neil, Thomas P., RM3

Ordway, Emerson L., CEM

Paulson, Roger F., S2c

Peel, John R., CMoMM

Pelliciari, Nicholas J., MoMM2

Pierce, Sam H., MoMM1

Pigman, Billy B., EM2

Pike, Neal, CRT

Potting, Roy C., Y3

Powers, Kenneth C., RM3

Rewold, Radford C., CMoMM

Riley, Thomas G., SC1c

Ring, Thomas G., SC1c

Sandleben, Francis J., CCS

Schmelzer, Carl T., CEM

Scholle, Donald J., QM3

Schuett, James S., SM1

Sergio, Michael, RM1

Shaw, Thomas E., EM3

Sigworth, Kenneth L., EM2

Taylor, John W., MoMM1

Terboss, William F., S1c

Ware, William J., Ens.

Wells, Francis A., CTM

Wood, Perry S., Lt.(jg)

Wright, Richard M., Lt.

PIERCE'S POLECATS

TUNNY

Adams, James W., S1c

Battles, Roy E., RM2

Baughman, Doyle F., SC2

Beaman, George B., TM3

Benjamin, Roland, S1c

Bress, Henry, Lt.

Bruce, William O., F2c

Busemeyer, Francis J., FCS2

Carini, Neno R., MoMM2

Carver, Robert J., MoMM1

Chase, Sheldon R., TM3

Chestnut, William J., FC1

Cisk, John S., EM3

Colegrove, Warren R., Ens.

Combs, Russell W., S1c

Corbeil, John W., Ens.

Dawson, Frank A., F1c

Delfino, James C., EM1

Doss, Richard H., S1c

Duda, Walter M., Bk2

Dunniny, Robert J., MoMM2

Eckels, Thomas J., S1c

Emerson, Lloyd D., GM3

Gannon, Thomas P., RM3

Garver, Jack E., S1c

Hagopian, Leon V., QM2

Harang, Richard A., Lt.(jg)

Hargrave, Daniel J., MoMM1

Heeney, William E., MoMM2

Hurt, Everett J., EM2

Hutchens, Robert C., MoMM2

Isaacson, Robert C., MoMM2

Johnson, Marvin J., MoMM3

Johnson, Woodrow W., PhoM3

Jones, Eugene C., CMoMM

Jones, Herbert E., Jr., Lt.(jg)

Jones, William A., S1c

Keiper, Arthur T., TM2c

Kimble, Floyd G., Jr., RM2

LaValley, Francis P., EM3

Leifhelm, Wilbur D., CMoMM

Levinson, Russell, Y1

MacDonald, Charles H., RT2

Mahaffa, Walt C., EM3

Martin, Everett E., MoMM2

Martin, Paul W., SC3

McCoy, Howard R., TM2

McMillin, George W., CQM

Merry, Everett E., MoMM2

Miller, Howard R., MoMM2

Mitchell, Patrick H., MoMM2

Negrette, Tony, RT1

Neher, Charles B., S2c

Nickisher, Rudolph F., EM3

Oppelt, Francis K., TM2

Owen, Edwin J., TM3

Peterson, Clarence G., EM3

Phoenix, Noel M., CMoMM

Pierce, George E., Cdr.

Potopinski, Arthur S., PhM1

Pulling, Thomas T., SM3

Reddinger, Edward D., S1c

Ritter, Charles H., EM2

Rouiller, Charles A., Lt.

Saffel, Elmer B., CEM

Sanderson, Dean H., S2c

Schonschack, Howard A., RT1

Sherman, Melvin C., S1c

Sibley, Richard O. B., S1c

Simpson, Charles E., QM1

Spafford, Robert T., F1c

Stone, William J., RT3

Teachman, David W., MoMM3

Thomas, Beachor, Ck3

Timmerman, Harold H., S1c

Trosper, Elmer E., CTM

Turner, James, StM2

Vaughn, Henry L., Lt. Cdr.

Voskuhl, Frederick H., EM2

Wade, John J., GM3

Waggoner, Richard Leon, TM3

Webber, William J., Jr., MoMM3

Weigant, Dean M., RM1

Zacharsuk, John, MoMM1

SKATE

Bailey, William H., EM3

Bauer, Frederick W., TM3

Bebeau, Walter H., S1c

Beste, John D., EM2

Bird, John J., EM1

Brennan, Philip F., MoMM3

Brown, George E., SC3

Bryson, Howard W., GM3

Burlin, Charles W., Jr., Lt.(jg)

Butler, Creighton F., CMoMM

Carlin, Thomas L., Lt.(jg)

Champion, Ralph A., Jr., S1c

Collier, John L., Bkr2

Covell, Wallace M., MoMM1

Coyne, Edward J., TM3

Crooks, Sheridan R., Lt.(jg)

Daniel, James, StM1

Davis, William M., S1c

Dearing, Harry J., MoMM1

Debuhr, Calvin H., Lt.(jg)

Donovan, John L., EM2

Doyle, Reginald E., WOE

Duzik, Emil F., EM2

Earhart, Herman M., Ens.

Eastwood, Freddie R., TM1c

Edgerton, Stuart T., Lt.

English, Cannon M., CMoMM

Ewald, Marcus H., S1c

Farnof, Arthur, Jr., S1c

Faurotte, Harvey R., TM1

Foster, Theodore, MoMM3

Galles, Lester D., MoMM1

Gann, Earnest E., F1c

Glabb, Richard G., RM3

Goss, Herbert W., TM2

Heissenbuttel, Samuel, EM3

Hermance, Frank J., MoMM1

Hill, Joseph T., TM2

Hinton, Floyd J., MoMM2

Huston, Robert C., Lt. Cdr.

Jordan, James I., F1c

Kenyon, Frank C., III, S1c

Kice, Everett F., EM1

Kichline, Reginald G., CMoMM

Knold, Vernon O., GM2

Ledbetter, Frank T., CRM

Lee, Arnold, S1c

Lemier, Billy B., SM2

Levy, Stanley S., S1c

Lynch, Richard B., Cdr.

Marshall, Douglas S., Jr., S1c

McCracken, Chadwick N., SC2

McFadyen, Peter J., QM2

Mileskie, Stanley J., EM2

Miller, Robert E., MoMM3

Millspaugh, Stanley C., MoMM3

Moller, Edwin R., RT2

Moser, Adam D., EM3

Mudore, Thomas A., MoMM1

Murphy, John D., S1c

Murphy, Robert L., Jr., EM3

Naylor, William R., Jr., S1c

O'Donohue, Robert J., S1c

Olufsen, Albert E., RM2

Ostrom, Norman H., MoMM3

Parker, Edgar, Jr., RM2

Paul, Clinton J., F1c

Perigo, Robert L., S1c

Potter, John R., FC2

Praskievicz, Wallace M., CEM

Rayner, William A., MoMM2

Roberts, Grady E., S1c

Ruediger, Manfred W., TM1

Schlotterer, Jack C., MoMM2

Shelton, William A., GM1

Smith, Frederick H., Jr., MoMM2

Smith, Jack C., RT1

Southwick, Paul, PhoM1

Spencer, Amos A., S1c

Talucci, Ralph M., S1c

Tandy, William H., Jr., PhM1

Thomas, Leroy, St3

Turbitt, William J., F1c

Wagoner, William B., S1c

West, Franklin G., Lt.

Whiting, George L., CY

Wiest, Irvin C., CTM

Wilkins, Kenneth M., RT3

Woods, Isaac B., Jr., EM1

BONEFISH

Abel, Donald A., Ens.

Adams, Thomas B., Y3

Adams, Wendell S., Bkr3

Amburgey, Lawrence M., Lt.(jg)

Anderson, Gustav I., Jr., MoMM3

Aueril, Sestilio J., S1c

Beck, Merle L., GM2

Brown, Roderick W., F1c

Browning, James A., EM1

Burdick, Charles A., MoMM2

Canfield, Kenneth T., MoMM2

Coleman, John A., RM3

Cooley, Quintus L., StM2

Danielson, Otis C., SC2

Dunn, Davis H., Ens.

Edge, Lawrence L., Cdr.

Enos, E. R., F1c

Epps, William H., Jr., StM2

Feld, Paul E., F1c

Fox, Donald C., RM2

Frank, Richard E., CMoMM

Fugett, Mack A., QM2

Fuller, Grant M., CMoMM

Hackstaff, Howard J., RM2

Harman, Guy P., TM1

Hasiak, John J., TM3

Hess, Richard D., S1c

Houghton, Wilbur S., TM1

Jenkins, Robert W., EM1

Johnson, John C., RT1

Johnson, Stuart E., Jr., CQM

Johnston, Russel M., Lt.(jg)

Kalinoff, Michael W., F1c

Karr, William G., RM2

Keefer, Robert T., S1c

Kern, Franklyn B., Ens.

King, Edward W., EM2

Kissane, John E., S2c

Knight, Fraser S., Lt. Cdr.

Lamothe, Joseph N., Cox

Laracy, John J., Jr., EM3

Lewis, Marion A., CGM

Lockwood, Thomas G., PhoM3

Lynch, Joseph F., TM2

Maghan, Allan G., F1c

Markle, John E., EM2

McBride, Roy J., MoMM2

Miles, Henry V., MoMM1

Nester, Sidney A., EM3

Newberry, Joseph R., F1c

Olson, Douglas H., MoMM2

O'Toole, William P., EM3

Parton, John F., EM3

Paskin, Theodore, RT2

Pauley, George W., RM3

Phenicie, John E., MoMM3

Primavera, Louis J., MoMM1

Prunier, George A., EM3

Quenett, Clayton F., TM2

Raley, Charles H., F1c

Ray, Roscoe G., Jr., SM1

Raynes, James A., EM1

Reid, Jack A., F1c

Rhanor, Charles J., S1c

Rice, Robert M., S1c

Rose, Russell A., Ens.

Schiller, Robert G., F1c

Schmidling, Charles J., FCS1

Schweyer, Roy H., RT2

Slater, Robert E., Lt.(jg)

Smith, Logan C., Lt.(jg)

Snodgrass, Roger L., Y1

Stamm, Raymond S., SC1

Surber, Robert M., EM2

Tierney, Daniel R., MoMM1

Velie, Russell C., TM2

Vincent, Thomas F., Jr., S1c

Whitright, Willard, TM2

Williams, Jay J., MoMM2

Williams, John R., Jr., F3c

Williams, Thomas F., F1c

Wilson, Joseph R., F1c

Winegar, Clarence D., TM3

Wolfe, Lynn E., TM3

Wright, George "W," PhM1

RISSER'S BOBCATS

FLYING FISH

Anderson, Lloyd C., CmoMM

Anthony, Melvin L., RT3

Apostolopoulos, Vasilios, QM3

Bartocci, Lawrence E., F1c

Beardslee, Ralph C., Jr., SM1

Bennett, Wilfred A., TM2

Birkner, Francis R., PhoM1

Burke, Julian T., Jr., Lt. Cdr.

Canaday, Gerald B., TM1

Caramenico, Lewis J., F1c

Cates, Don B., FC1

Chereek, Benjamin, EM1

Christensen, Charles R., CMoMM

Collins, Joel W., III, RM2

Cooper, Earl B., MoMM2

Cronin, Joseph J., MoMM1

Doheny, Edward L., III, Lt.

Drozdowicz, Edward J., MM1

Dunn, Matthew D., EM2

Early, John A., Jr., F1c

Emmons, Robert C., WOE

Evans, Cassel J., TM2

Fiedler, John E., F2c

Field, Sidney F., F1c

Funkhouser, Edward M., RM3

Geiser, Robert F., RM3

Giannelli, Frank A., MoMM3

Griffin, Charles W., SC3

Hall, Lloyd A., RT1

Haney, Arthur A., RT2

Hayes, William, EM3

Herbert, Edward R., F1c

Holland, Noble V., Bkr1

Holloway, Harold E., S2c

Holzwarth, Jacob T., MoMM2

Hopley, Eric E., Lt.

Jasinski, Leon F., EM3

Jerbert, Arthur H., Lt.(jg)

Johnson, Kenneth W., EM3

Kenworthy, Harvey W., F1c

Kilgore, William H., Lt.

Kocon, Joseph S., TM2

Korn, Carl A., Jr., Lt.

Laster, Robert, StM1

Logan, John E., GM1

Lort, Joseph M., Jr., QM2

Lusse, Melvin R., EM2

Lynsky, Mark V., Jr., MoMM2

Mahoney, Robert C., TM1

Mattingly, John W., TM1

McGee, John S., Y1

Miller, Kay D., F1c

Moody, Dick, MoMM2

Nelson, Chester R., F1c

Nelson, William C., S1c

Nickerson, Bryan W., S1c

O'Brien, William J., FCS2

Ostergren, John F., Lt.

Peterman, John J., Jr., TM2

Ragsdale, Glenn E., S1c

Rainer, William H. Jr., RM3

Rankin, Walter H., MoMM2c

Risser, Robert D., Cdr.

Rodgers, James L., S1c

Rusin, Nicholas, TM3

Schmersahl, Jacob B., Jr., MoMM3

Schoomaker, Edward P., S1c

Schulke, Oscar J., EM2

Shaw, Harold M., RM2

Sly, Richard H., Ens.

Smith, Billy R., SC1

Smith, Carl, MoMM3

Smith, Paul T., S1c

Sproull, Raymond D., Jr., S1c

Sunbury, George G., RT3

Thacker, William O., StM1

Thompson, Earnest L., MoMM1

Wakshinsky, Albert S., EM2

Ward, James M., MoMM2

Weeks, Richard L., S1c

Whitefield, William B., CPhM

Wildes, Warren F., EM3

Wilson, Robert G., F1c

Witt, Ishmael C., MoMM2

BOWFIN

Alexander, James R., Bkr3

Alexander, Robert E., CPhM

Alpin, Carter F., Jr., EM3

Anderson, Hubert C., Lt.

Ayers, James M., Ens.

Beales, Austin W., MoMM1

Benson, Gordon H., RM2

Beyer, Walter L., QM1

Beynon, Robert P., EM3

Bruderly, Robert E., EM2

Buckman, Horace T., F1c

Carberry, Jack S., F1c

Carden, Olie L., MoMM1

Carter, Arthur L., FC1c

Chisum, Albert, TM3

Choquette, Hugh E., TM3

Clarey, John L., Lt. Cdr.

Cole, William E., TM1

Cummins, William E., Lt.(jg)

Curran, John E., MoMM2

Elliott, Michael M., Lt.(jg)

Ely, John A., S1c

Erickson, John H., EM1

Ervin, Norval L., Jr., S1c

Flessner, Conrad J., Lt.

Fletcher, Earl T., EM3

Gaito, Eugene, CMoMM

Gilkes, Thomas H., MoMM1

Gillespie, Clark H., MoMM1

Gilmore, Paul D., MoMM3

Gosnell, Marshall S., MoMM1

Harrington, John L., MoMM2

Hedland, Fred, F1c

Heinz, Robert L., Y1

Holder, John "A," F1c

Howard, Homer L., F2c

Jackson, Ted D., F1c

Jackson, Wilbur L., MoMM1

Johnson, Gerald B., EM2

Johnson, Henry N., S1c

Kear, Charles B., F1c

Kenney, Albert P., Bkr3

King, Henry, Jr., TM1

Knoche, Eugene A., MoMM2

Knox, Joseph M., Jr., CMoMM

LaCour, Marshall W., PhoM2

Lancaster, Wallace E., S2c

Launius, John J., RT1

Lundgren, Walter E., MoMM3

McMillion, George J., F1c

McNeven, Vern, MoMM2

Molloy, Leslie R., RT2

Nash, Paul G., GM2

Odoms, Edward A., StM1

Ohlund, Arley "V," S1c

Olsen, Rolf S., EM2

Patterson, Robert G., CEM

Perske, Earl W., CCS

Poppleton, Sidney R., S1c

Price, Lloyd R., S1c

Rasp, Vincent R., F1c

Reiner, Morton M., S1c

Rodskiaer, Aage E., RM1

Rohrbacher, Virgel H., GM1

Ryan, Ronald R., S1c

Scaglione, Peter T., RM3

Stack, Thomas P., TM3

Sweat, James G., SM1

Taylor, Richard S., EM2

Turner, Charles R., StM2

Tyree, Alexander K., CDR

Updegraff, Jack L., QM3

Van Kuran, Peter, Lt.(jg)

Verkinder, Victor, MoMM1

Videkovich, William P., MoMM2

Waddell, Kenneth A., EM2

Waugh, William L., TM2

Weidner, Alpheus S., Jr., MoMM3

Weller, Homer G., QM2

Winning, Edward G., CEM

Wise, John P., Ens.

TINOSA

Anderson, Jack H., EM3

Atnip, Tolbert B., S1c

Ault, Earl E., F1c

Baird, Floyd C., MoMM1

Barr, George, MoMM3

Bennett, Millard M., TM3

Bentham, Robert E., TM3

Bolinder, Ralph H., F1c

Boyd, Edgar, SM1

Brady, Ferris G., RT3

Brooks, F. C., Lt.

Brumfeld, Floyd E., S1c

Burke, Charles M., MoMM3

Burlew, Harry A., MoMM3

Carlen, Robert C., S1c

Carpenter, Clarence A., S1c

Carpenter, Rex N., S1c

Clement, William R., TM2

Clutterham, D. R., Ens.

Costibile, Joe F., S1c

Daranowich, Walter H., PhoM3

Daughtry, Herbert, MoMM2

Dismukes, Alvin C., FCS2

Dixon, Richard L., TM1

Dowler, Melvin L., QM2

Eterovich, Matthew, F1c

Freeburn, Harry D., EM2

Garner, Frank E., PhM1

Gibson, Jack R., EM2

Giltner, Thomas W., MoMM1

Goen, Louis E., TM3

Gould, Harold Ray, S1c

Grigg, John R., MoMM3

Grose, H. G., Lt.(jg)

Groves, Russell C., RM1

Hall, William "F.," MoMM3

Harris, Frederick B., Y1

Hinds, Lawrence P., RM2

Huson, Loyal A., TM3

Irvin, Frederick L., SC3

Jackson, Nathaniel, StM2

Keepers, Harold J., S1c

Klag, Donald J., EM3

Larson, Allen G., F1c

Latham, R. C., Cdr.

Leonard, Carthel F., EM3

MacPherson, Malcolm, RM2

McDaniel, Jessie J., EM3

Minor, James P., EM2

Nylander, Raymond E., S1c

Olsen, J., Ens.

Otis, Donald J., TM2

Owens, Robert L., EM3

Paquette, Clifford N., F1c

Polis, Jack S., EM3

Reif, Harry W., MoMM2

Richeson, Edward M., GM1

Robbins, Kermit E., MoMM1

Robertson, James C., MoMM3

Rodman, George W., QM3

Salisbury, G. F., Lt.(jg)

Sanders, Aubrey R., S1c

Sanders, C. R., Lt.

Scott, Dale V., S1c

Scruggs, Robert C., SC1

Searles, Dayton, Jr., GM3

Settle, Spaulding B., ST3

Shelden, James F., S1c

Siegfried, C. W., Lt.(jg)

Smith, H. J., Lt. Cdr.

Smith, Richard P., MoMM3

Soutiere, Clement, EM3

Stamant, Wilfred J., Jr., S1c

Stanford, Samuel E., MoMM3

Stevens, Charles R., MoMM3

Stokes, Victor L., CEM

Stripling, Ernest R., MoMM2

Thompson, Robert M., MoMM2

Tyler, John P., RT1

Vannatter, Charles H., RM3

Voegtin, E. P., EM1

Wagner, Charles H., Jr., TM3

Weaver, B. S., Lt.(jg)

Welch, Freeman, CTM

Whipps, Jack C., RT3

Wicker, William A., QM3

Wilson, Eldon R., TM3

Wilson, Norval D., CmoMM

Young, Buck R., FC2

NOTES

Introduction

1   Blair, Clay, Jr., *Silent Victory*, Lippincott, 1975, Vol. 2, p. 851 (Hereafter, Blair); and Roscoe, Theodore, *United States Submarine Operations in World War II*, United States Naval Institute, 1949, pp. 491–92. (Hereafter, Roscoe.)

Chapter One: A World Destroyed

1   Telegram to Sarah S. Edge from U.S. Navy Bureau of Naval Personnel. Sarah Edge Shuler and Lawrence L. Edge Jr. (Hereafter, Edge family.)
2   Lawrence L. Edge to Sarah S. Edge. Edge family.
3   Ibid.
4   Ibid.

Chapter Two: ComSubPac

1   Some sixty years later the plane's wreckage reappeared after heavy rains washed away the covering soil.
2   Blair, p. 341.

Chapter Three: The *Wahoo*'s Last Dive

1   Roscoe, p. 240.
2   The *Squalus* flooded and sank in waters off the coast of New Hampshire on May 23, 1939, with the loss of twenty-six men. She was raised, overhauled, and recommissioned as the *Sailfish* on May 15, 1940.
3   Top-secret war patrol report of the USS *Lapon* (SS-260).
4   Top-secret war patrol report of the USS *Narwhal* (SS-167).

5 Top secret war patrol report of the USS *Sawfish* (SS-270).

6 Article included in a report submitted by Lockwood to Nimitz concerning the loss of the USS *Wahoo* (SS-238). Article and report attached to war patrol report of the *Wahoo*.

7 Lockwood, Charles A., and Hans C. Adamson, *Hellcats of the Sea*, New York: Greenburg, 1951, p. 2. (Hereafter, Lockwood, *HOS*.)

8 O'Kane, Richard H., *Wahoo: The Patrols of America's Most Famous WWII Submarine*, Novato, California: Presidio Press, 1987, p. 329.

9 Lockwood, *HOS*, p. 22.

Chapter Four: The Commander from Georgia

1 Roscoe, p. 531.

2 Lawrence L. Edge to Sarah S. Edge. Edge family.

3 Ibid.

4 Top secret war patrol report of the USS *Bonefish* (SS-223).

5 Ibid.

Chapter Five: The "Magic" Behind the Mission

1 FM Sonar, Model 1, Nos. 11–15, Preliminary Instruction Book, Operation and Maintenance. Unclassified. May 12, 1945 [Rpt. R223.4], Box 1. University of California Division of War Research at the U.S. Navy Radio and Sound Laboratory, San Diego, California. Courtesy of the Scripps Institution of Oceanography Archives, University of California, San Diego, La Jolla, California. (Hereafter, UCDWR.)

2 "FM Sonar," by M. C. Henderson and Charles A. Hisserich. Unclassified. September 4, 1943, Box 2. The conceptual basis for using FM sonar to locate mines underwater. UCDWR.

3 Lockwood, *HOS*, p. 38.

4 Lockwood, *HOS*, p. 42.

5 Sources concerning this meeting are at variance as to the actual date as well as the participants. King may have arrived in Pearl Harbor on or about July 12 or 13. He then departed on an inspection trip with Nimitz and other Pacific commanders (there is no mention of Lockwood), minus MacArthur, to the Marianas. After this tour King apparently returned to Washington before Roosevelt's arrival in Honolulu on July 26. Yet according to Lockwood, his meeting with King and Nimitz took place on July 26, five days *after* King's departure and the day of Roosevelt's arrival.

Chapter Six: Wolf Pack

1   Top-secret war patrol report of the USS *Bonefish*.
2   Ibid.
3   Ibid.
4   Ibid.
5   Copy of original Navy message to *Bonefish* from ComTwelve. Edge family.

Chapter Seven: The Long Road to Tokyo

1   Top-secret prospectus for Operation Barney; Lockwood to Nimitz and King, December 3, 1944, "Japan Sea—Patrol of." National Archives II (NARA II), Modern Military Records, College Park, MD. RG 313. 5.3 Records of Naval Operating Forces, including those of Operation Barney. Courtesy of Steven Trent Smith.
2   Lockwood had waived the ironclad rule that subs on patrol never, ever broadcast when operating in an area where Japanese radio-direction-finding stations could triangulate their position. In practice the rule was often violated whenever submarine skippers sent contact reports concerning enemy convoys, task forces, and other important information to ComSubPac and to other submarines that, if patrolling nearby, could provide more information or join in an attack.

Chapter Eight: The Magic Loses Its Magic

1   Lockwood to Lawrence L. Edge. November 13, 1944. Edge family.
2   In summer 1943, Watkins, then a submarine division commander, asked Lockwood for permission to make a war patrol as a skipper. The USS *Flying Fish* (SS-229) was available, and Watkins, at age forty-five, became the oldest man to ever command a sub on a war patrol. Watkins's performance was mediocre at best. Ironically, Lockwood had requested permission from Nimitz to make a war patrol, not as a skipper but as a rider, and was turned down. Nimitz felt it was too risky to put his sub force commander in a position to be captured and tortured for information from the Japanese.
3   Lockwood to Rear Adm. C. W. Styer, USN. Lockwood papers, Box 14. Library of Congress. (Hereafter, LPLC.)
4   Lockwood to Watkins, January 26, 1945. LPLC, Box 14.
5   Nimitz to Lockwood, *HOS*, p. 67.
6   Lawrence L. Edge to Mr. and Mrs. Ralph W. Edge (Lawrence Edge's parents). Edge family.

Chapter Nine: An Operation Called "Barney"

1   Lockwood to Vice Admiral R. S. Edwards, USN. LPLC, Box 15.
2   Lockwood, *HOS*, p. 71.
3   Lockwood to Captain H. C. Bruton, USN. LPLC, Box 15.

Chapter Ten: The Minehunters

1   Lawrence L. Edge to Sarah S. Edge. April (undated) 1945. Edge family.
2   Ibid. Edge family.
3   Top-secret war patrol report of the USS *Bonefish*.
4   Lawrence L. Edge to Sarah S. Edge. Edge family.
5   Ibid.

Chapter Eleven: Probing the Line

1   Top-secret war patrol report of the USS *Seahorse* (SS-304).
2   Ibid.
3   Ibid.
4   Ibid.
5   Lockwood, *HOS*, p. 88.
6   Top-secret war patrol report of the USS *Seahorse*.
7   Top-secret war patrol report of the USS *Bonefish*.
8   Ibid.
9   Ibid. Endorsement letter by Edge's division commander, Captain Louis Chappell, USN.
10 Lawrence L. Edge to Sarah S. Edge. Edge family.
11 Ibid.

Chapter Twelve: "Hydeman's Hellcats"

1   Letter from Lawrence L. Edge to Jane Tharpe, May 10, 1945. Edge family.
2   Lawrence L. Edge to Sarah S. Edge. Edge family.
3   Ibid. Edge family.
4   Unpublished manuscript by Earl T. Hydeman concerning submarine operations and Operation Barney. Undated. Courtesy of Barbara Hydeman Barnes. (Hereafter, Hydeman Ms.)
5   Lawrence L. Edge to his parents. Edge family.
6   Lockwood, *HOS*, p. 110. Neither Voge nor Lockwood explained how this arrangement was to be worked out with the Russians, as they were supposed to be kept in the dark about U.S. subs in the Sea of Japan. Perhaps they simply hoped that if a U.S. sub showed up at Vladivostok, they'd be welcomed for the twenty-

four-hour time limit imposed upon warships seeking refuge in neutral ports. Likely Lockwood just kept his fingers crossed that nothing would happen that would require porting there.

7   ComSubPac Operation Order No. 112-45, May 26. Declassified. NARA II. Modern Military Records, College Park, MD. RG 313. 5.3 Records of Naval Operating Forces, including those of Operation Barney. (Hereafter, Op ord 112-45.)

8   Lockwood, *HOS*, p. 114. The original SORG document defied attempts to locate it at NARA II and the LPLC. Lockwood makes reference to it in both *HOS* and in his memoir, *Sink 'Em All*.

Chapter Thirteen: Running the Gauntlet

1   Commander Earl T. Hydeman's Standing Orders Log for the period May 27–July 4, 1945. Courtesy of Robert Barry and Patricia Hydeman Barry.

2   Top-secret addendum to the war patrol report of the USS *Sea Dog*.

3   "War and Remembrance: The Mighty Mine Dodgers; Saga of the *Sea Dog*, Sea of Japan, June 4–25, 1945." From SubmarineSailor.com Internet posting, August 1, 1998. www.submarinesailor.com

4   Top-secret addendum to the war patrol report of the USS *Sea Dog*.

5   Ibid.

6   Top-secret addendum to the war patrol report of the USS *Spadefish*.

Chapter Fourteen: Threading the Needle

1   Top-secret war patrol report of the USS *Skate* (SS-305).

2   From a description of the *Tinosa*'s penetration of the Tsushima Strait told to Lockwood by skipper Latham (*HOS*, pp. 142–47). The incident is described in a single short paragraph in Latham's top secret addendum to the *Tinosa*'s Sea of Japan patrol report. The description gives no details other than the fact that the mine cable made contact with the hull outside the after engine room. In fact, Latham reported to Lockwood that the contact originated outside the hull at about the conning tower, which is slightly forward of the middle of the ship. In his patrol report Latham merely stated the bare fact that "This [noise] is believed to have been a mine cable from the [FMS] contact [of a mine] on the port bow."

3   Lockwood to Watkins, June 1, 1945. LPLC, Box 15.

4   Watkins to Lockwood, June 9, 1945. Ibid.

Chapter Fifteen: The Death of an Empire

1   Latham's comment on the sinking of the *Wakatama Maru*, published in the April 1981 *Tinosa Blatt* newsletter.

2   Though Lockwood said that the torpedo problem had been solved by late 1944,

it's clear from reading the patrol reports of the Hellcats (and other subs on patrol late in the war) that it wasn't. There were still far too many erratic runs (see the *Tinosa*'s experience with one of her own torpedoes that made a circular run), broachers, and duds. The same problems that had bedeviled the Mk 14 and Mk 18 torpedoes were evident in the newer Mk 23s. Due mainly to these problems the Hellcats sank fewer ships than they could have. That failure can also be traced in part to poorly executed attacks, overeagerness, and faulty judgment on the part of the Hellcat skippers. While the Hellcats took a sizable toll in ships sunk, many got away unscathed.

3   Top-secret report of the USS *Crevalle* (SS-291).

4   Top-secret patrol report of the USS *Skate* (SS-305).

5   Ibid.

6   Top-secret patrol report of the USS *Flying Fish* (SS-229).

7   Ibid.

8   In *Silent Victory* (p. 839), Clay Blair says that Lockwood radioed the Hellcats, "Did anybody shoot northwest of La Pérouse Strait?" The answer came back from Germershausen, who suspected he'd erred: "Guilty." When the Hellcats returned after their mission Germershausen received a summons to Nimitz's office. Questioned by the admiral, the skipper told his side of the story and was told by Nimitz, "Glad you made it back safely, son." The episode reinforced Lockwood's determination to keep Soviet warships, especially submarines, from operating in the Sea of Japan. By then Nimitz had dropped the idea altogether.

Chapter Sixteen: A Dark Silence

1   Hydeman Ms.

2   There is no evidence that the *Crevalle*'s crew killed the Japanese sailor in question. However, the fact that they tried to corresponds to the one issue that, despite Dudley W. Morton's outstanding war record, has, in some critics' view, left him tarnished. That issue is the gun attack Morton ordered unleashed on the survivors of a troop transport sunk by the *Wahoo* on January 26, 1943. In his patrol report Morton described battle surfacing among the hundreds of survivors (some said thousands), many of them in so-called "troop boats," or lifeboats, of various kinds. When the *Wahoo*'s gunners started shooting at them, their fire was returned by what Morton described as "[S]mall caliber machineguns. We then opened fire with everything we had" (USS *Wahoo* (SS-238), third patrol report, p. 58). In Morton's judgment this apparently made the Japanese survivors fair game, as he went on to mow them down. If submarine command had misgivings about Morton's actions it's not apparent in the glowing endorsements to his patrol report, one of which reads, "An outstanding patrol. This patrol speaks for itself, and the judgment and decisions [of the commanding officer] demonstrate what can be done by a submarine that retains the initiative." It's interesting

to note that the International Military Tribunal for the Far East—the Japanese war crimes trials—brought charges that the Japanese regularly machine-gunned survivors of Allied ships that had been sunk as well as Allied POW survivors of Japanese slave ships sunk by U.S. forces. (See USS *Bonefish* sinking of POW ship, footnote p. 49–50.

During Operation Barney the *Spadefish*, *Flying Fish*, *Tinosa*, and *Bowfin* also attacked and sank small craft with their guns, but their patrol reports contain no mention of gun crews shooting at survivors in the water.

3 *Bowfin* patrol report.

4 Top secret patrol report of the USS *Tunny* (SS-282).

5 Neither Pierce's nor Lynch's patrol report makes any reference to hearing explosions coming from Toyama Wan.

Chapter Seventeen: Breakout

1 Lockwood, June 23, 1945. LPLC, Box 15.

2 Lockwood, *HOS*, p. 294.

3 Lockwood to James Fife, June 27, 1945. LPLC, Box 15.

4 Lockwood to Nimitz, July 18, 1945. Ibid.

5 Ibid.

6 Lockwood diary. LPLC, Box 1.

Chapter Eighteen: The Long Search

1 Sarah S. Edge to Lawrence L. Edge, July 26, 1945. A handwritten note on the letter says, "Never mailed as Gov't telegram came three days later." This is a reference to the missing-in-action telegram Sarah received on July 28, 1945. Edge family.

2 Beach, Edward L., *Submarine!* New York: Henry Holt, 1946.

3 *Atlanta Journal*, August 12, 1945. "Son Born Day After Skipper Announced Lost."

4 Ibid.

5 *Los Angeles Times*, August 12, 1945, p. 4. "Sub Flotilla Returns After Taking Nip Toll."

6 Lockwood to Sarah S. Edge, August 12, 1945. Edge family.

7 Lucius H. Chappel to Sarah S. Edge, August 12, 1945. Edge family.

8 Potter, E. B., *Nimitz*, Annapolis, Maryland: Naval Institute Press, 1976, p. 388. In the context of the situation of August 11, King's message prefix could almost be taken as a rejoinder to the infamous "This is a war warning" message transmitted to the hapless Admiral Kimmel before the Japanese attack on Pearl Harbor.

9 Lockwood to Sarah S. Edge, September 13, 1945. Edge family. Lockwood does not make clear in his reply to Sarah if, in referring to the "*Bonefish* going on this special mission at her own request," he meant Operation Barney or entry into

Toyama Wan. In the letter, Lockwood makes reference to Edge's seeking permission from Pierce to enter the bay, which under the circumstances was the proper thing to do. It would make sense that if Lockwood understood that Sarah was seeking clarification of the latter point, he would forward her letter to Pierce, because as force commander the admiral knew that Operation Barney was not organized on a volunteer basis. Because few family members knew anything at all about the *Bonefish*'s operations in Toyama Wan, it seems likely that any questions they had about volunteering referred to Operation Barney itself.

10 Ibid.

Chapter Nineteen: The Hour of Sacrifice

1 George Pierce to Sarah S. Edge, September 18, 1945. Edge family.

2 The questions Sarah posed show her extraordinary grasp of the tactical situation as it pertained to Operation Barney. It's not clear from any of the extant correspondence between her and Lockwood or the Department of the Navy how she acquired this information, as none of Lockwood's letters nor those from McCann, Chappell, and others go into the tactical details of Operation Barney. One suspects that there may have been other newspaper or magazine articles, which she saw, that did.

3 Sarah S. Edge to Richard B. Lynch, October 3, 1945. Edge family. The letter as quoted is a rough draft with many excisions. Typically, Sarah produced rough drafts of her correspondence, and this is one of only a few that survive in the Edge family archive.

4 Richard B. Lynch to Sarah S. Edge, October 16, 1945. Edge family.

5 The so-called "families" letter is undated. It contains all the information she'd received from Lynch and others as noted. She makes no references to comments made by some family members of the crew that the men had done more than their fair share and that, having been ordered to undertake Operation Barney and not receiving more help when they needed it, the order was akin to murder. For the full text of the letter see Appendix Three. Edge family.

Chapter Twenty: A Shining Glory

1 Allen R. McCann to Sarah S. Edge, June 21, 1946. Edge family.

INDEX

Peter Sasgen is the author of two additional books on submarine warfare, *Stalking the Red Bear: The True Story of a U.S. Cold War Submarine's Covert Operations Against the Soviet Union* and *Red Scorpion: The War Patrols of the USS* Rasher. He lives in Florida.